The Body Atlas

A pictorial guide to the human body

Illustrated by Giuliano Fornarni

Written by Steve Parker

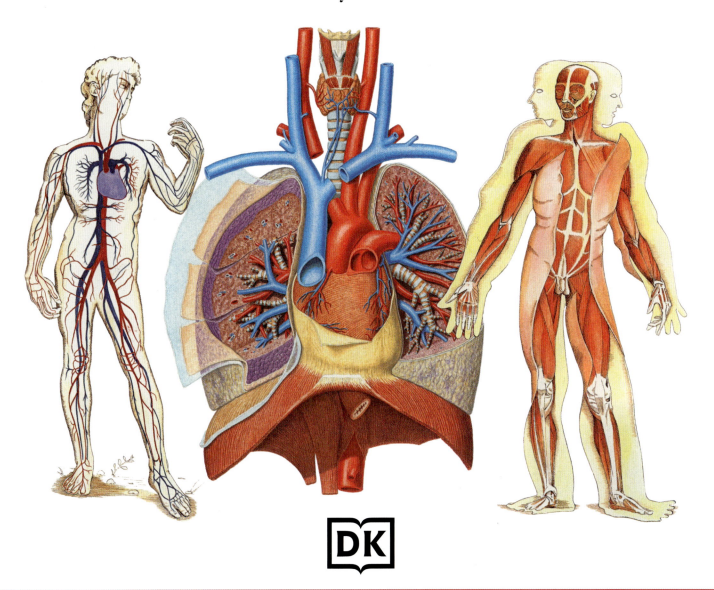

DK

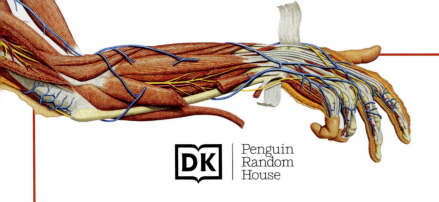

DK | Penguin Random House

REVISED EDITION
Senior Editors Abigail Mitchell, Virien Chopra
Senior Art Editor Vikas Chauhan
Editor Ben Ffrancon Davies
Art Editor Sanya Jain
Managing Editors Christine Stroyan, Kingshuk Ghoshal
Managing Art Editors Anna Hall, Govind Mittal
Picture Researcher Nimesh Agrawal
Production Editor Kavita Varma
Senior Production Controller Jude Crozier
Senior DTP Designer Vishal Bhatia
Jacket Designer Akiko Kato
Jacket Design Development Manager Sophia MTT
Publisher Andrew Macintyre
Art Director Karen Self
Publishing Director Jonathan Metcalf
Medical Consultant Dr Kristina Routh MBChB, MPH

FIRST EDITION
Senior Art Editor Christopher Gillingwater
Project Editors Laura Buller, Constance Novis
Designer Dorian Spencer Davies
Managing Editor Susan Peach
Managing Art Editor Jacquie Gulliver
Medical Consultant Dr Thomas Kramer MBBS, MRCS, LRCP

This edition published in 2020
First published in Great Britain in 1993 by
Dorling Kindersley Limited
DK, One Embassy Gardens, 8 Viaduct Gardens,
London, SW11 7BW

Copyright © 1993, 2018, 2020 Dorling Kindersley Limited
A Penguin Random House Company
10 9 8 7 6 5 4 3 2 1
001–316675–Sep/2020

A CIP catalogue record for this book is
available from the British Library.
ISBN: 978-0-2414-1277-0

Printed and bound in China

For the curious
www.dk.com

CONTENTS

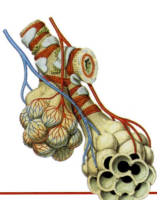

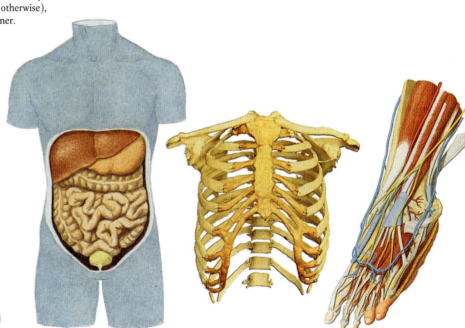

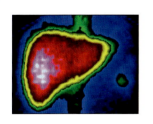  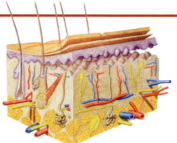 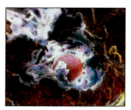 

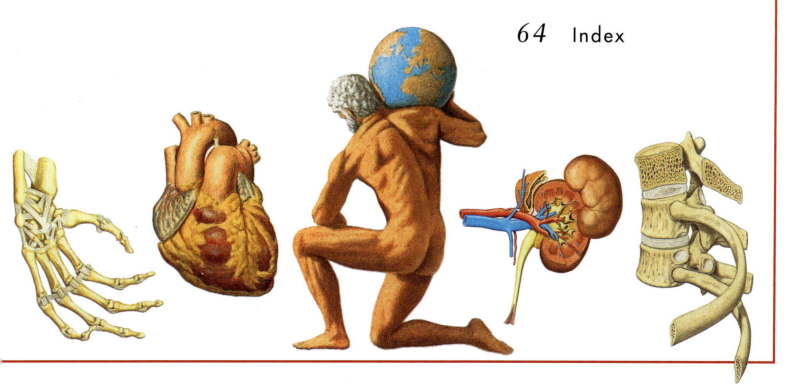

Mapping the Human Body

THIS BOOK INTRODUCES the exciting world inside you – the world of the human body. The drawings on the following pages explore the body from head to toe, one major region at a time. The artworks give you a closer look at these vital structures, showing you how each organ works and interacts with the body parts around it.

Your body may not look exactly like the body mapped here. Where there are differences – as there are between the female and male sexes, for example – separate artworks show you what these differences are. There are also variations in height, weight, skin colour, hair, bone proportions, and other features, that make each of us an individual within the human species *Homo sapiens*.

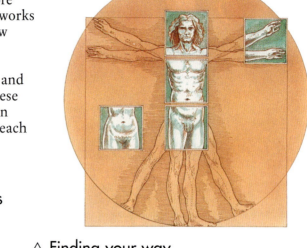

◁ The outsides of your insides

The *Body Atlas* shows what your body parts look like from the outside and the inside – even from the front and back. Some body parts are shown in realistic form and colour, as in this view of the outside of the heart, its wrappings, and the major blood vessels surrounding it.

△ Finding your way

The *Body Atlas* is divided into sections that deal with the major regions of the body, such as the head and neck, the upper torso, or the arm and hand. Each section is identified next to the main page title by an illustration of that region, as shown above. These illustrations provide a quick guide to the location of the region in the body.

▷ Inside your insides

Body parts are also shown with the outsides peeled back to reveal the insides. This view, for example, shows the inner chambers, linings, valves, and muscles of the heart. Arteries, veins, and nerves are shown in three different colours.

LAYER BY LAYER

This book lifts away each layer of the body, so that you can see how the structures inside fit and work with those around them. The body is covered with a layer of flexible skin, under which are layers of fat and muscle. Further in lie the bones of the skeleton which provide a firm framework for support and movement. The main organs of the body, such as the brain, lungs, heart, liver, stomach, and intestines, lie deep within the body, protected by bone and muscle.

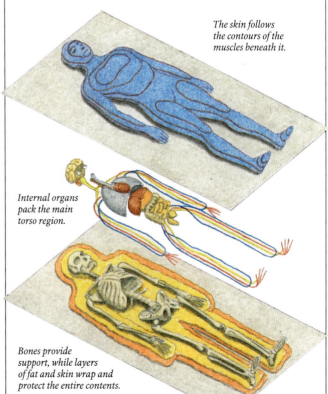

The skin follows the contours of the muscles beneath it.

Internal organs pack the main torso region.

Bones provide support, while layers of fat and skin wrap and protect the entire contents.

▽ Images from life

Photographs in the *Body Atlas* use modern imaging techniques, so you can examine parts you could never see with your eyes. Colour-enhanced scans and X-rays, such as the one on the right, reveal the bones and organs under the skin. Below is a photo taken with a scanning electron microscope. It displays structures as small as an individual cell.

A micrograph (a photograph taken with a microscope) of a glomerulus, the ball of tiny capillaries inside the kidney

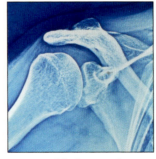

An X-ray of the bones in the shoulder joint, showing how the rounded end of the upper-arm bone fits into a cup-shaped socket in the shoulder blade

NAMING NAMES

Most parts of the body have Latin names. Some of these names may seem strange at first, but once you learn their meanings, you will see how helpful they are:

Superficialis	Towards the outside or surface
Profundus	Towards the inside or underneath
Anterior	Towards the front
Posterior	Towards the back
Dorsal	Nearer the back or top
Ventral	Nearer the front, belly, or bottom
Superior	Above, bigger, or more important
Inferior	Below, smaller, or less important
Medial	Towards the middle or midline

▽ Under the skin

This is an example of the fascinating illustrations found over the following pages. The skin of the arm and hand has been lifted away so that you can see the muscles, bones, nerves, and blood vessels beneath, all precisely labelled.

A section of this muscle has been left out to show what is underneath.

This ligament is cut away so you can see the tendons it secures.

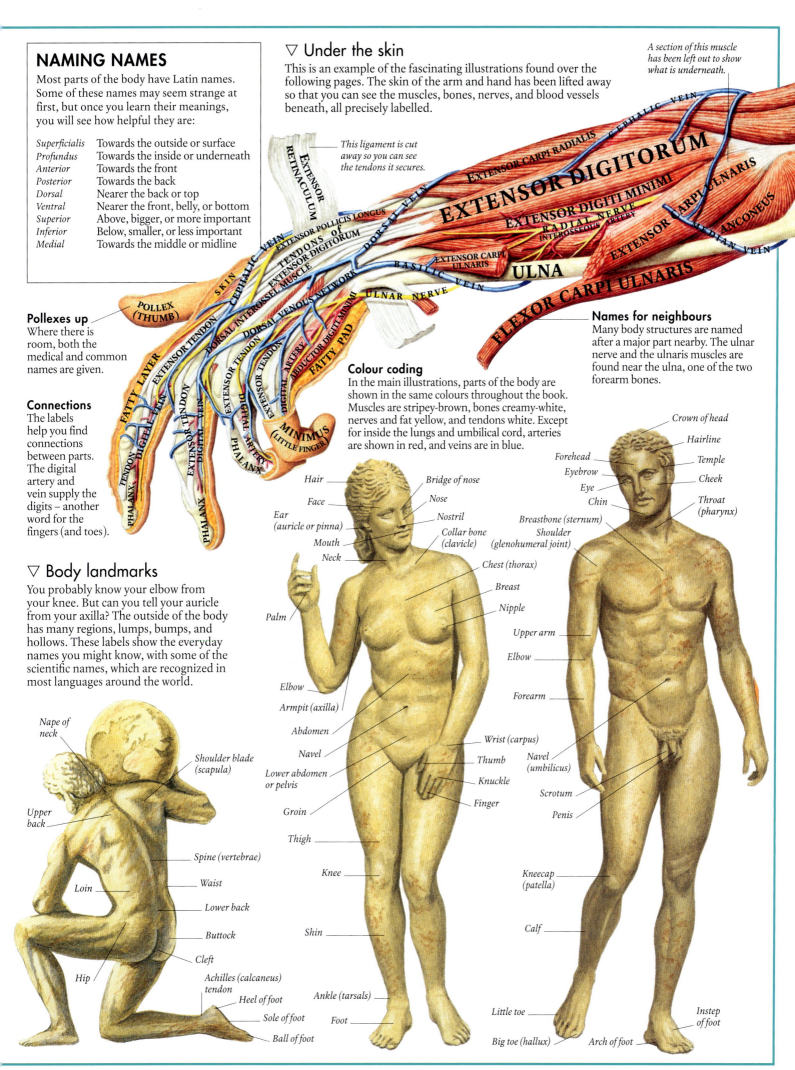

EXTENSOR DIGITORUM

EXTENSOR CARPI RADIALIS

CEPHALIC VEIN

EXTENSOR DIGITI MINIMI

EXTENSOR CARPI ULNARIS

RADIAL NERVE

INTEROSSEOUS ARTERY

ANCONEUS

MEDIAN VEIN

EXTENSOR CARPI ULNARIS

ULNA

EXTENSOR RETINACULUM

EXTENSOR POLLICIS LONGUS

TENDONS of EXTENSOR DIGITORUM

DORSAL VEIN

EXTENSOR DIGITORUM MUSCLE

BASILIC VEIN

FLEXOR CARPI ULNARIS

POLLEX (THUMB)

SKIN

CEPHALIC VEIN

FATTY LAYER

DORSAL INTEROSSEL NETWORK

DORSAL VENOUS NETWORK

DORSAL ARTERY DIGITI MINIMI

ABDUCTOR DIGITI MINIMI

ULNAR NERVE

FATTY PAD

EXTENSOR TENDON

DIGITAL VEIN

DIGITAL ARTERY

TENDON

PHALANX

EXTENSOR TENDON

DIGITAL VEIN

EXTENSOR TENDON

PHALANX

EXTENSOR TENDON

DIGITAL ARTERY

PHALANX

MINIMUS (LITTLE FINGER)

Pollexes up
Where there is room, both the medical and common names are given.

Connections
The labels help you find connections between parts. The digital artery and vein supply the digits – another word for the fingers (and toes).

Names for neighbours
Many body structures are named after a major part nearby. The ulnar nerve and the ulnaris muscles are found near the ulna, one of the two forearm bones.

Colour coding
In the main illustrations, parts of the body are shown in the same colours throughout the book. Muscles are stripey-brown, bones creamy-white, nerves and fat yellow, and tendons white. Except for inside the lungs and umbilical cord, arteries are shown in red, and veins are in blue.

▽ Body landmarks

You probably know your elbow from your knee. But can you tell your auricle from your axilla? The outside of the body has many regions, lumps, bumps, and hollows. These labels show the everyday names you might know, with some of the scientific names, which are recognized in most languages around the world.

Nape of neck

Shoulder blade (scapula)

Upper back

Loin

Hip

Spine (vertebrae)

Waist

Lower back

Buttock

Cleft

Achilles (calcaneus) tendon

Heel of foot

Sole of foot

Ball of foot

Hair

Face

Ear (auricle or pinna)

Mouth

Neck

Palm

Elbow

Armpit (axilla)

Abdomen

Navel

Lower abdomen or pelvis

Groin

Thigh

Knee

Shin

Ankle (tarsals)

Foot

Bridge of nose

Nose

Nostril

Collar bone (clavicle)

Chest (thorax)

Breast

Nipple

Wrist (carpus)

Thumb

Knuckle

Finger

Crown of head

Hairline

Forehead

Temple

Eyebrow

Cheek

Eye

Chin

Throat (pharynx)

Breastbone (sternum)

Shoulder (glenohumeral joint)

Upper arm

Elbow

Forearm

Navel (umbilicus)

Scrotum

Penis

Kneecap (patella)

Calf

Little toe

Big toe (hallux)

Arch of foot

Instep of foot

Body Systems

THE TRILLIONS OF CELLS inside your body link to form tissues – special cell groups with one main job. Groups of one or more kinds of tissues make up your body's main parts, called the organs. A collection of organs that work together to carry out a particular task is called a body system. You are reading this book, for example, using your eyes, which are part of your sensory system. They are scanning the illustrations and words and, in an instant, reporting the information to your brain, the central organ of your nervous system.

At the same time, other body systems are carrying out the tasks that keep you alive. Your heart and blood vessels form your circulatory system, which carries blood around your body, and your lungs are the main part of your respiratory system, which breathes in air.

Each system has a distinct function, but they all work together to keep your whole body functioning smoothly and efficiently.

The body's main systems

Your major body systems are shown on the next four pages. In some systems, the parts are grouped together. For example, most of the organs of the digestive system are packed into your abdomen. In others, such as the circulatory system, the parts – in this case, the blood vessels – are spread throughout the body.

INSIDE A CELL

Cells are microscopic building blocks. More than 37 trillion of them make up your bones, muscles, nerves, skin, blood, and other organs and body tissues. The drawing below shows a "typical" cell, cut away to show the even smaller parts, called organelles, inside it. Cells similar in shape to the one shown here exist only in a few parts of the body, such as the liver. In most other parts, the cells are of different sizes and shapes, specialized to do particular jobs. Other specialized cells are the nerve cells above, the fat cell opposite, and the blood cells on page 8.

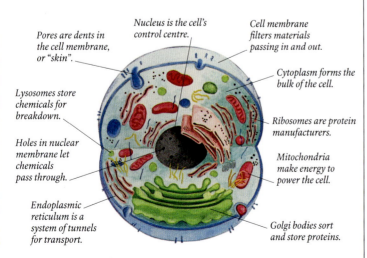

Pores are dents in the cell membrane, or "skin".

Nucleus is the cell's control centre.

Cell membrane filters materials passing in and out.

Lysosomes store chemicals for breakdown.

Cytoplasm forms the bulk of the cell.

Holes in nuclear membrane let chemicals pass through.

Ribosomes are protein manufacturers.

Mitochondria make energy to power the cell.

Endoplasmic reticulum is a system of tunnels for transport.

Golgi bodies sort and store proteins.

▽ Skeletal system

From head to toe, your bones play a vital role supporting the softer parts of your body. Your skeleton consists of about 206 bones, linked to each other at joints. They provide a strong, yet flexible framework that is moved by your muscles. Some bones surround and protect more delicate organs. The skull shields your brain, and the ribs encase your heart and lungs.

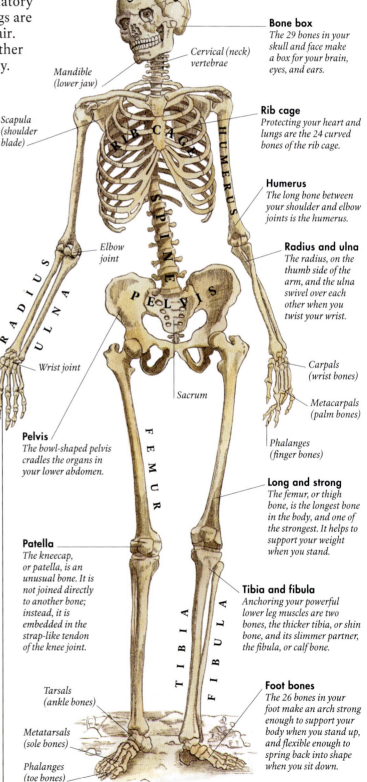

Mandible (lower jaw)

Cervical (neck) vertebrae

Scapula (shoulder blade)

Elbow joint

Wrist joint

Pelvis
The bowl-shaped pelvis cradles the organs in your lower abdomen.

Sacrum

Patella
The kneecap, or patella, is an unusual bone. It is not joined directly to another bone; instead, it is embedded in the strap-like tendon of the knee joint.

Tarsals (ankle bones)

Metatarsals (sole bones)

Phalanges (toe bones)

Bone box
The 29 bones in your skull and face make a box for your brain, eyes, and ears.

Rib cage
Protecting your heart and lungs are the 24 curved bones of the rib cage.

Humerus
The long bone between your shoulder and elbow joints is the humerus.

Radius and ulna
The radius, on the thumb side of the arm, and the ulna swivel over each other when you twist your wrist.

Carpals (wrist bones)

Metacarpals (palm bones)

Phalanges (finger bones)

Long and strong
The femur, or thigh bone, is the longest bone in the body, and one of the strongest. It helps to support your weight when you stand.

Tibia and fibula
Anchoring your powerful lower leg muscles are two bones, the thicker tibia, or shin bone, and its slimmer partner, the fibula, or calf bone.

Foot bones
The 26 bones in your foot make an arch strong enough to support your body when you stand up, and flexible enough to spring back into shape when you sit down.

▽ Muscular system

There are more than 600 muscles in your body, providing the pulling power so that you can move about. Most are attached to bones or to other muscles by tough cords called tendons. Many body organs, such as the heart, intestines, and bladder, contain their own muscle. Most muscles are only known by their scientific names, which often twist the most flexible muscle of all – your tongue!

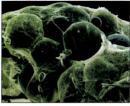

A fat, or adipose, cell contains "blobs" of fat that is used for energy when food is scarce.

▽ Integumentary system

This is the name for the skin, hair, nails, and other parts covering your body. The surface of the skin consists of dead, hardened cells like tiles on a roof, that are rubbed off as you move and wash. Yet just below the surface, your skin is very much alive. Its cells multiply every second to replace those worn away.

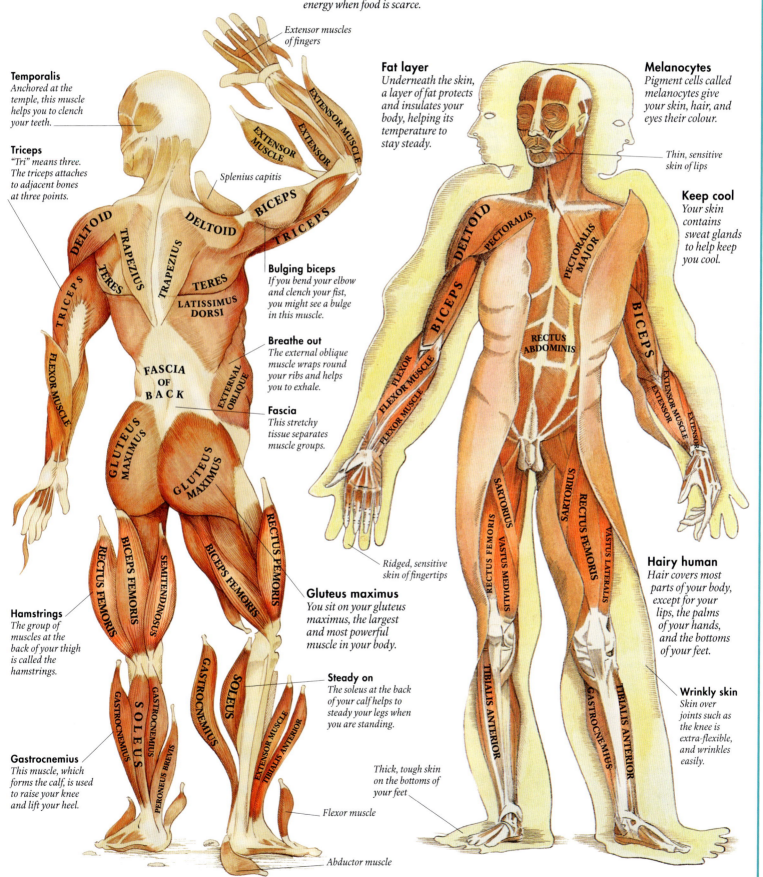

Temporalis
Anchored at the temple, this muscle helps you to clench your teeth.

Triceps
"Tri" means three. The triceps attaches to adjacent bones at three points.

Extensor muscles of fingers

Splenius capitis

Fat layer
Underneath the skin, a layer of fat protects and insulates your body, helping its temperature to stay steady.

Melanocytes
Pigment cells called melanocytes give your skin, hair, and eyes their colour.

Thin, sensitive skin of lips

Keep cool
Your skin contains sweat glands to help keep you cool.

Bulging biceps
If you bend your elbow and clench your fist, you might see a bulge in this muscle.

Breathe out
The external oblique muscle wraps round your ribs and helps you to exhale.

Fascia
This stretchy tissue separates muscle groups.

Ridged, sensitive skin of fingertips

Gluteus maximus
You sit on your gluteus maximus, the largest and most powerful muscle in your body.

Steady on
The soleus at the back of your calf helps to steady your legs when you are standing.

Hamstrings
The group of muscles at the back of your thigh is called the hamstrings.

Gastrocnemius
This muscle, which forms the calf, is used to raise your knee and lift your heel.

Flexor muscle

Abductor muscle

Hairy human
Hair covers most parts of your body, except for your lips, the palms of your hands, and the bottoms of your feet.

Wrinkly skin
Skin over joints such as the knee is extra-flexible, and wrinkles easily.

Thick, tough skin on the bottoms of your feet

Body Systems (*Continued*)

YOUR BODY NEEDS A constant supply of energy. Your digestive and respiratory systems pass fuel and oxygen to your blood; the circulatory system, helped by the lymphatic system, delivers these substances to cells, and takes away wastes for removal by the excretory system. Co-ordinating all these actions are the nervous and endocrine systems.

▷ Nervous and endocrine systems

The nervous system connects your entire body with the brain. Sensory nerves carry signals from your sense organs to your brain; once the brain decides what to do, it sends messages to your muscles along motor nerves. This system works closely with the endocrine system, which uses chemical messengers called hormones to control many body processes, as well as growth and development.

Sensory signals
Your brain gets information from outside of your body from the organs of your main senses – sight, hearing, smell, taste, touch, and balance.

Eye and optic nerve for sight

BRAIN

Pituitary
Found just below the brain, this gland releases hormones to control other glands.

Thyroid
This gland helps to regulate body growth.

SPINAL CORD

THORACIC NERVES

PANCREAS

Adrenal glands
When you are angry or frightened, these glands tell your body to get ready for action.

LUMBAR NERVES

SACRAL NERVES

A SENSE OF TASTE

Your tongue is a mobile muscle that enables you to taste food, move it around as you chew, and push it back into your throat when swallowing, and speak up to ask for more. Its rough surface is covered with tiny bumps called papillae. On and between these are microscopic onion-shaped bunches of cells – the taste buds. These detect flavours and send signals to the taste centre in your brain. Your taste buds can detect five basic tastes: sweet, salty, bitter, sour, and umami (savoury).

Epiglottis covers your windpipe when you swallow food.

EPIGLOTTIS

PALATINE TONSIL

LINGUAL TONSIL

PALATINE TONSIL

Lingual and palatine tonsils contain germ-killing cells.

APEX OF TONGUE

All areas of the tongue, especially around the sides and the back, have taste buds, with each detecting one of the five tastes.

Motor nerves control the muscles of the lower body.

Touch signals
Senses are more complex than they seem. The sense of touch in your fingertips involves pressure, heat, cold, and pain.

Toe nerves
Sensory nerves stretch right into your toe-tips, telling you when you have stubbed your toe.

▽ Circulatory system

Fresh, oxygen-rich blood pumped from the heart reaches all parts of your body through arteries. These branch into tiny capillaries, where blood gives oxygen and nutrients to the surroundings, and takes in wastes for disposal. Veins collect the used blood and take it back to the heart.

Carotid artery
Blood reaches your brain through this artery.

Aorta
The aorta carries fresh blood from the heart to the rest of your body.

Jugular vein
Deep in the neck, this vein returns blood from the head to the heart.

HEART

Heart
This muscular, two-part pump keeps blood circulating around your body.

Vena cava
Used blood from your organs pours into the vena cava for the return journey to your heart.

Femoral artery to leg

Extra push
The massaging movements of the calf muscles help to pump blood in the lower legs upwards, against the pull of gravity.

Perforating vein in calf muscles

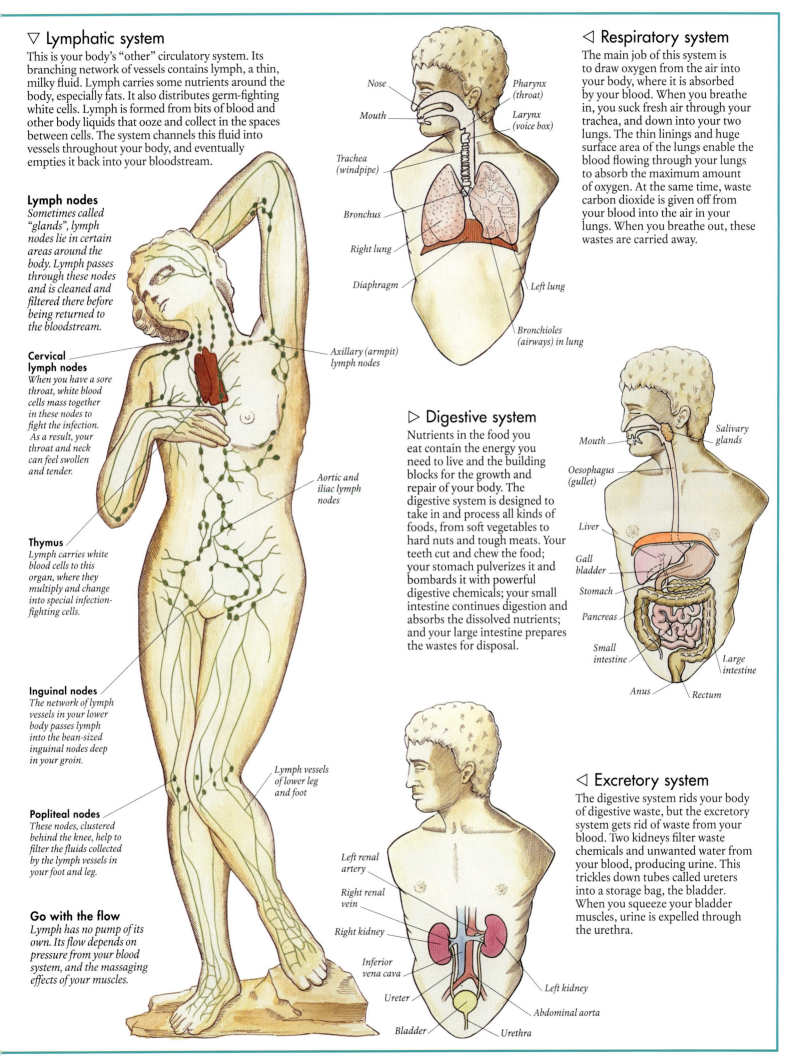

▽ Lymphatic system

This is your body's "other" circulatory system. Its branching network of vessels contains lymph, a thin, milky fluid. Lymph carries some nutrients around the body, especially fats. It also distributes germ-fighting white cells. Lymph is formed from bits of blood and other body liquids that ooze and collect in the spaces between cells. The system channels this fluid into vessels throughout your body, and eventually empties it back into your bloodstream.

Lymph nodes

Sometimes called "glands", lymph nodes lie in certain areas around the body. Lymph passes through these nodes and is cleaned and filtered there before being returned to the bloodstream.

Cervical lymph nodes

When you have a sore throat, white blood cells mass together in these nodes to fight the infection. As a result, your throat and neck can feel swollen and tender.

Thymus

Lymph carries white blood cells to this organ, where they multiply and change into special infection-fighting cells.

Inguinal nodes

The network of lymph vessels in your lower body passes lymph into the bean-sized inguinal nodes deep in your groin.

Popliteal nodes

These nodes, clustered behind the knee, help to filter the fluids collected by the lymph vessels in your foot and leg.

Go with the flow

Lymph has no pump of its own. Its flow depends on pressure from your blood system, and the massaging effects of your muscles.

Nose
Mouth
Trachea (windpipe)
Bronchus
Right lung
Diaphragm
Pharynx (throat)
Larynx (voice box)
Left lung
Bronchioles (airways) in lung

Axillary (armpit) lymph nodes
Aortic and iliac lymph nodes
Lymph vessels of lower leg and foot

◁ Respiratory system

The main job of this system is to draw oxygen from the air into your body, where it is absorbed by your blood. When you breathe in, you suck fresh air through your trachea, and down into your two lungs. The thin linings and huge surface area of the lungs enable the blood flowing through your lungs to absorb the maximum amount of oxygen. At the same time, waste carbon dioxide is given off from your blood into the air in your lungs. When you breathe out, these wastes are carried away.

▷ Digestive system

Nutrients in the food you eat contain the energy you need to live and the building blocks for the growth and repair of your body. The digestive system is designed to take in and process all kinds of foods, from soft vegetables to hard nuts and tough meats. Your teeth cut and chew the food; your stomach pulverizes it and bombards it with powerful digestive chemicals; your small intestine continues digestion and absorbs the dissolved nutrients; and your large intestine prepares the wastes for disposal.

Mouth
Oesophagus (gullet)
Liver
Gall bladder
Stomach
Pancreas
Small intestine
Salivary glands
Large intestine
Anus
Rectum

◁ Excretory system

The digestive system rids your body of digestive waste, but the excretory system gets rid of waste from your blood. Two kidneys filter waste chemicals and unwanted water from your blood, producing urine. This trickles down tubes called ureters into a storage bag, the bladder. When you squeeze your bladder muscles, urine is expelled through the urethra.

Left renal artery
Right renal vein
Right kidney
Inferior vena cava
Ureter
Bladder
Left kidney
Abdominal aorta
Urethra

The Head and Neck

SITTING ON THE TOP of your body, balanced on your neck, is the head. Just under the surface lies the skull, a strong, bony structure that houses the brain – the body's nerve centre. This wrinkled mass of cells lets you speak, think, and learn. Your senses of smell, hearing, taste, and sight are located here.

Peel back the skin, fat, and muscles of your head and neck, and this is what you would see. The front of the head is a mass of about 30 muscles that control the eyes, face, and mouth. Below the main part of the head are the muscles that bend and twist the neck, and the blood vessels that link the head with the heart.

▷ Going for the jugular

The carotid arteries carry blood to the head and brain, while the jugular veins bring it back to the heart. In humans and other animals, these vessels are near the surface of the neck. When an animal catches its prey, it may go for the throat to rip open these vessels. Vital blood supply to the brain is cut off, and so much blood pours out that the prey dies fast.

Muscle sheet
The occipitofrontalis is a sheet of muscle that wraps around your skull. It helps to wrinkle the forehead, raise the eyebrows, and pull back the scalp.

Nose muscle
The nasalis muscle in your nose has two parts. When you flare your nostrils, you use the alar part, and when you wrinkle your nose, you use the transverse part.

Facial vein
The facial vein carries blood from your upper face, down the side of your nose and under your eye.

VOCAL CORDS

The vocal cords, used for speaking and singing, are located in your throat, within the larynx, or voice box (right). This is made of cartilage and muscle tissue. The cords are made of strips of tough, elastic tissue, covered by a delicate membrane. Muscles in the larynx move the cords, stretching them loosely for low sounds and tightly for high ones. These photographs show the vocal cords as a doctor sees them, looking down the throat with a laryngoscope.

Cords pulled tightly together for speech

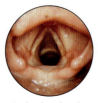

Cords relaxed and apart for breathing

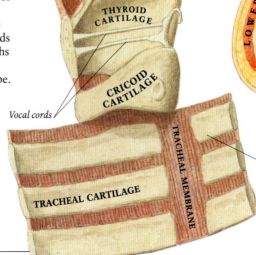

Vocal cords

HYOID BONE
EPIGLOTTIS
MEMBRANE
THYROID CARTILAGE
CRICOID CARTILAGE
TRACHEAL CARTILAGE
TRACHEAL MEMBRANE

Air passes from the lungs through the trachea, or windpipe, shown here sliced open and laid flat. This air makes the vocal cords vibrate, producing the sound of your voice.

FAT LAYER UNDER THE SKIN
OCCIPITOFRONTALIS
DURA
ARACH
PIA MATER CO
SKULL
ORBICULARIS OCULI (EYELID MUSCLE)
EYE
FACIAL VEIN
FACIAL VEIN
NASALIS MUSCLE
FACIAL ARTERY
SMALL CHEEK MUSCLE
CHEEKBONE
LARGE CHEEK MUSCLE
NASAL CAVITY
HARD PALATE
LOWER JAW
UPPER LIP
TONGUE
LOWER LIP
LOWER JAW
GENIOGLOSSUS MUSCLE (ROOT OF THE TONGUE)
MASSETER MUSCLE
LOWER JAW MUSCLE
HYOID BONE
NECK MUSCLE
NECK MUSCLE

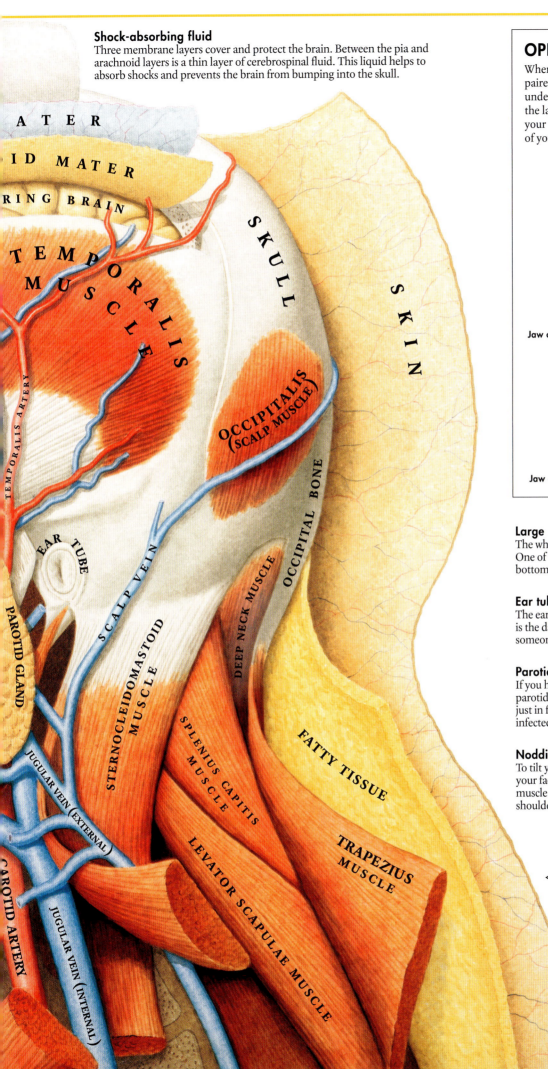

Shock-absorbing fluid
Three membrane layers cover and protect the brain. Between the pia and arachnoid layers is a thin layer of cerebrospinal fluid. This liquid helps to absorb shocks and prevents the brain from bumping into the skull.

ATER

ID MATER

RING BRAIN

TEMPORALIS MUSCLE

SKULL

SKIN

TEMPORALIS ARTERY

OCCIPITALIS (SCALP MUSCLE)

OCCIPITAL BONE

EAR TUBE

SCALP VEIN

PAROTID GLAND

STERNOCLEIDOMASTOID MUSCLE

DEEP NECK MUSCLE

SPLENIUS CAPITIS MUSCLE

FATTY TISSUE

JUGULAR VEIN (EXTERNAL)

LEVATOR SCAPULAE MUSCLE

TRAPEZIUS MUSCLE

AROTID ARTERY

JUGULAR VEIN (INTERNAL)

OPEN WIDE OR SHUT UP!
Whenever you chat or chew, three sets of paired muscles are hard at work. The muscles under your jaw pull your mouth shut, while the lateral pterygoid muscle pulls the rear of your lower jaw forwards. This tilts the front of your lower jaw down, opening your mouth.

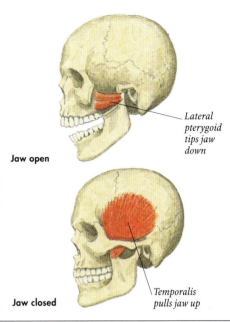

Lateral pterygoid tips jaw down

Jaw open

Temporalis pulls jaw up

Jaw closed

Large skull bone
The whole skull is a complex jigsaw of 22 bones. One of the largest is the occipital bone, at the bottom of the rear of the skull.

Ear tube
The ear tube, also known as the outer ear canal, is the dark hole you see when you look down someone's ear. It funnels sounds to the eardrum.

Parotid gland
If you have had mumps, you know where your parotid gland is! This saliva-producing gland, just in front of your ear, swells painfully when infected by mumps virus.

Nodding and shaking
To tilt your head from side to side, or to turn your face, you use the sternocleidomastoid muscle. When you look up or shrug your shoulders, you are using the trapezius muscle.

◁ **Hold your head up**
A complex set of intertwined muscles supports and moves your head, neck, and shoulders. Your head weighs about 5 kg (11 lb) and it takes a lot of effort to hold it up. The muscles in a new baby's neck are not fully developed, so a baby's head must be constantly supported, until the muscles become strong enough.

The Scalp and Skull

UNDER THE SURFACE OF your face lies a network of muscles, blood vessels, nerves, and sense organs, all arranged around the solid skull.

The main part of the skull is made of 22 separate bones, which join together during childhood forming wiggly lines called sutures. Eight of these bones form a protective box called the calvaria, around the brain. A further 14 bones, from the delicate lacrimal bone at the inner corner of your eye socket to the powerful mandible bone of the lower jaw, give your face its shape. The skull also contains the smallest bones in your body – three tiny bones in your ears called ossicles.

The scalp is a specialized area of skin that covers much of the top, sides, and rear of your head. It contains about 100,000 tiny hair follicles, from which hairs grow.

▽ The bones of the skull

This side view shows the separate bones of the skull. The skull bones cradle and protect the main sense organs as well as the brain. The temporal bone contains the delicate parts of the inner ear. Six bones on each side of the nose make a deep bowl, called the eye socket. The olfactory organs, which you use to smell, are in the nasal cavity behind the two small nasal bones.

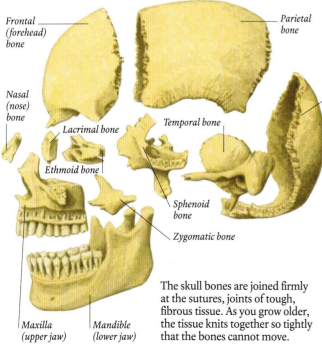

Frontal (forehead) bone

Parietal bone

Nasal (nose) bone

Lacrimal bone

Ethmoid bone

Temporal bone

Occipital bone

Sphenoid bone

Zygomatic bone

Maxilla (upper jaw)

Mandible (lower jaw)

The skull bones are joined firmly at the sutures, joints of tough, fibrous tissue. As you grow older, the tissue knits together so tightly that the bones cannot move.

Facial veins

A system of veins carries blood down from the centre top of the scalp to empty into the facial vein. The skin covering your head and face has a lot of blood vessels, which is why cuts there bleed so much.

Blood vessels

A network of small arteries, mostly sandwiched between the skin and the underlying muscles, supplies blood to your face and scalp. When you feel hot or embarrassed, blood rushes into these arteries to help the extra body heat escape. As a result, your face turns red for a moment.

Thin skin

Your skin is not the same thickness all over your body. The thinnest is the skin covering your eyelids.

TEMPORALIS MUSCLE

FRONTALIS MUSCLE

ORBICULARIS OCULI (EYELID MUSCLE)

MIDDLE TEMPORAL VEIN

TEMPORAL ARTERY

ZYGOMATIC

TRANSVERSE FACIAL ARTERY

SUPERFICIAL TEMPORAL VEIN

ZYGOMATICUS MAJOR

SKIN

FACIAL VEIN

MASSETER

SUPERIOR LABIAL ARTERY

BUCCINATOR

FAT LAYER

FACIAL ARTERY

INFERIOR LABIAL ARTERY

EXTERNAL CAROTID ARTERY

INFERIOR LABIAL VEIN

LINGUAL ARTERY

Under the skin

Just below your skin is a layer of fat, which helps to slow heat loss so that your body temperature is less affected by cold weather.

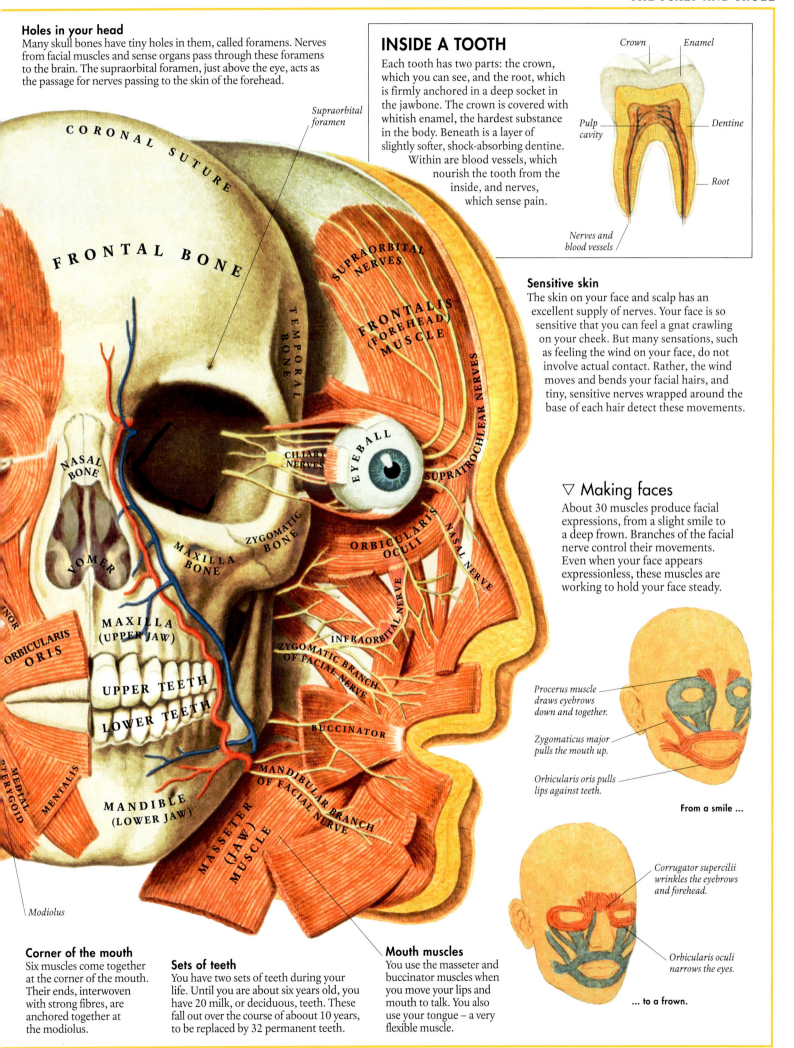

Holes in your head

Many skull bones have tiny holes in them, called foramens. Nerves from facial muscles and sense organs pass through these foramens to the brain. The supraorbital foramen, just above the eye, acts as the passage for nerves passing to the skin of the forehead.

INSIDE A TOOTH

Each tooth has two parts: the crown, which you can see, and the root, which is firmly anchored in a deep socket in the jawbone. The crown is covered with whitish enamel, the hardest substance in the body. Beneath is a layer of slightly softer, shock-absorbing dentine. Within are blood vessels, which nourish the tooth from the inside, and nerves, which sense pain.

Crown
Enamel
Pulp cavity
Dentine
Root
Nerves and blood vessels

CORONAL SUTURE

FRONTAL BONE

Supraorbital foramen

SUPRAORBITAL NERVES

FRONTALIS (FOREHEAD) MUSCLE

TEMPORAL BONE

SUPRATROCHLEAR NERVES

CILIARY NERVES

EYEBALL

NASAL BONE

ZYGOMATIC BONE

ORBICULARIS OCULI

NASAL NERVE

VOMER

MAXILLA BONE

SUPRAORBITAL NERVE

INFRAORBITAL NERVE

MAXILLA (UPPER JAW)

ZYGOMATIC BRANCH OF FACIAL NERVE

ORBICULARIS ORIS

UPPER TEETH

LOWER TEETH

BUCCINATOR

MENTALIS

MANDIBULAR BRANCH OF FACIAL NERVE

MEDIAL PTERYGOID

MANDIBLE (LOWER JAW)

MASSETER (JAW) MUSCLE

Modiolus

Sensitive skin

The skin on your face and scalp has an excellent supply of nerves. Your face is so sensitive that you can feel a gnat crawling on your cheek. But many sensations, such as feeling the wind on your face, do not involve actual contact. Rather, the wind moves and bends your facial hairs, and tiny, sensitive nerves wrapped around the base of each hair detect these movements.

▽ Making faces

About 30 muscles produce facial expressions, from a slight smile to a deep frown. Branches of the facial nerve control their movements. Even when your face appears expressionless, these muscles are working to hold your face steady.

Procerus muscle draws eyebrows down and together.

Zygomaticus major pulls the mouth up.

Orbicularis oris pulls lips against teeth.

From a smile ...

Corrugator supercilii wrinkles the eyebrows and forehead.

Orbicularis oculi narrows the eyes.

... to a frown.

Corner of the mouth

Six muscles come together at the corner of the mouth. Their ends, interwoven with strong fibres, are anchored together at the modiolus.

Sets of teeth

You have two sets of teeth during your life. Until you are about six years old, you have 20 milk, or deciduous, teeth. These fall out over the course of about 10 years, to be replaced by 32 permanent teeth.

Mouth muscles

You use the masseter and buccinator muscles when you move your lips and mouth to talk. You also use your tongue – a very flexible muscle.

The Brain

YOUR BRAIN IS THE nerve centre of your body. It controls most of your body movements, and gathers and stores information so that you can think and learn. Your brain weighs just 1.3 kg (3 lbs) with the consistency of stiff jelly, and more wrinkles than a giant walnut. It rests inside your upper skull within the calvaria, a "box" of bones. There it is safely supported and protected from knocks and jolts. For extra protection, three thin layers of membrane, the meninges, lie like a triple-decker sandwich between your brain and your skull bones.

The cerebrum makes up nine-tenths of your brain. Most of your thoughts, feelings, and emotions occur within this mass of nerve cells. The cerebrum is divided into two rounded halves, known as cerebral hemispheres. These two halves are joined by a "bridge" of nerve fibres, called the corpus callosum. The remaining tenth of your brain, located under the cerebrum, includes the cerebellum, pons, and medulla. These merge into the top of the spinal cord.

Lobes

Several sulci (grooves) divide each cerebral hemisphere into five main areas, or lobes. These are the prefrontal, frontal, parietal (at the top), temporal (at the side), and occipital (at the back). Each lobe has its own group of mental functions, although some functions are carried out by several lobes.

OCCIPITOFRONTALIS (MUSCLE SHEET)

FRONTAL SINUS

Frontal sinus

The sinuses are air-filled cavities within the thick skull bones, joined by openings to the main airway inside the nose. When you have a cold, your sinuses may fill up with mucus, and you suffer that "stuffed up" feeling.

NASAL CAVITY

▷ Nerve centre

The brain sends and receives messages through nerves that run down into the spinal cord. There are also 12 pairs of nerves that join directly to the brain and branch out into the head and neck. These are known as cranial nerves. One pair, the first cranial or olfactory nerve, is for smelling; the second cranial or optic nerve extends from the eye and is used for seeing. The fifth pair, the trigeminal nerve, has branches to the face, scalp, nose, mouth, and teeth.

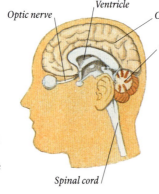

Optic nerve
Ventricle
Cerebrum
Cerebellum
Spinal cord

At the optic chiasma, near the hypothalamus, the optic nerves divide, and cross and combine.

OPTIC CHIASMA

Anterior part makes hormones.

PITUITARY STALK

ANTERIOR LOBE

BLOOD VESSEL NETWORK

BLOOD VESSEL NETWORK

POSTERIOR LOBE

Pituitary stalk connects pituitary to hypothalamus.

Blood vessels in the stalk carry hormones and chemicals to and from the pituitary.

Posterior part of pituitary gland makes and stores hormones.

GREY MATTER

When you think about something, you are using the outer layers of your cerebral hemispheres, called the cerebral cortex. The cerebral cortex also analyzes signals from your senses. The outermost layer is 3 mm (0.1 in) of "grey matter", consisting of nerve cells and their short, interwoven branches. Beneath it is the white matter, mainly bundles of long fibres that link the parts of the brain to each other.

Scan through a cross-section of the brain

Skull
White matter
Scalp
Grey matter

▷ Nerve-hormone connection

The pituitary gland makes hormones to control growth and to regulate activity in other glands and organs. It receives instructions, in the form of hormone-like chemicals, from the hypothalamus, a sugar-cube sized network of nerves which monitor the levels of hormones and chemicals in the body.

▷ Brain map

On the outside, each cerebral hemisphere looks the same all over. In fact, it is divided into several different areas or centres, each with its own job. Sensory centres receive nerve signals from your senses. Motor centres send signals out to your body muscles. The main body motor centre is subdivided into parts that control your tongue, lips, face, fingers, and other body regions.

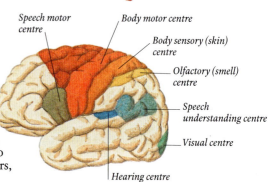

Speech motor centre
Body motor centre
Body sensory (skin) centre
Olfactory (smell) centre
Speech understanding centre
Visual centre
Hearing centre

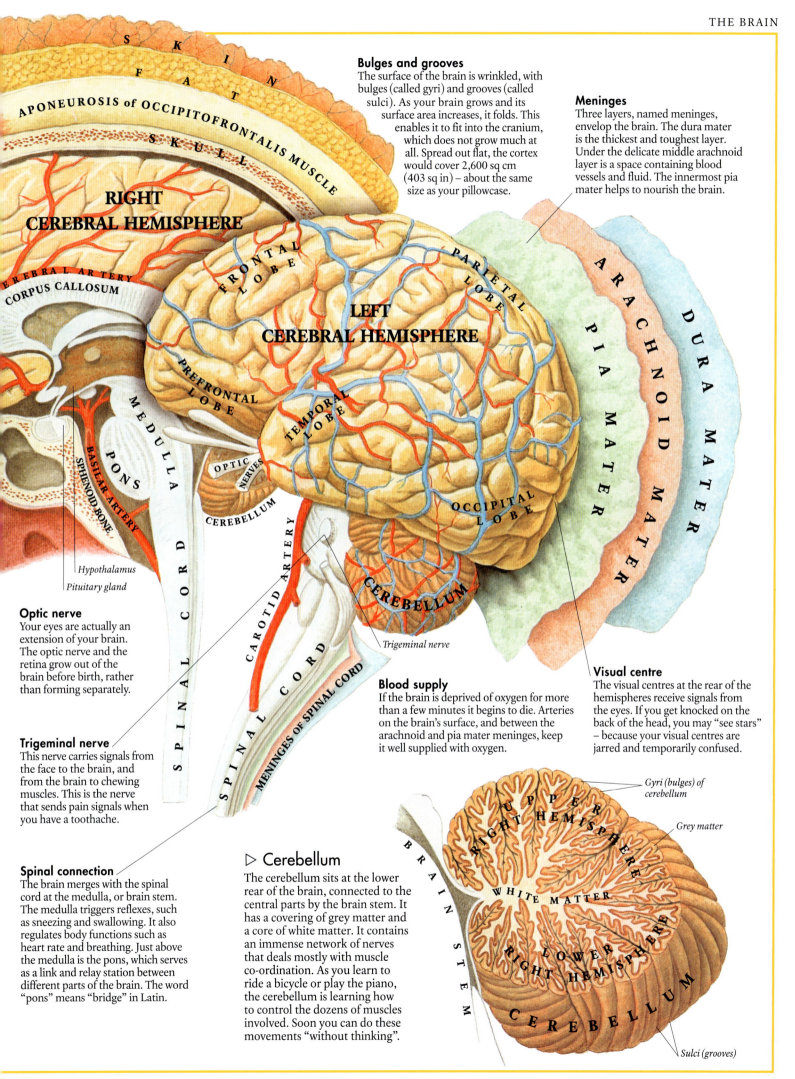

S K I N
F A T
APONEUROSIS of OCCIPITOFRONTALIS MUSCLE
S K U L L

RIGHT CEREBRAL HEMISPHERE

CEREBRAL ARTERY
CORPUS CALLOSUM

FRONTAL LOBE

LEFT CEREBRAL HEMISPHERE

PARIETAL LOBE

PREFRONTAL LOBE

TEMPORAL LOBE

M E D U L L A

P O N S

BASILAR ARTERY
SPHENOID BONE

OPTIC NERVES

CEREBELLUM

Hypothalamus
Pituitary gland

OCCIPITAL LOBE

CEREBELLUM

S P I N A L C O R D

CAROTID ARTERY

S P I N A L C O R D

MENINGES OF SPINAL CORD

Trigeminal nerve

A R A C H N O I D M A T E R

P I A M A T E R

D U R A M A T E R

Bulges and grooves
The surface of the brain is wrinkled, with bulges (called gyri) and grooves (called sulci). As your brain grows and its surface area increases, it folds. This enables it to fit into the cranium, which does not grow much at all. Spread out flat, the cortex would cover 2,600 sq cm (403 sq in) – about the same size as your pillowcase.

Meninges
Three layers, named meninges, envelop the brain. The dura mater is the thickest and toughest layer. Under the delicate middle arachnoid layer is a space containing blood vessels and fluid. The innermost pia mater helps to nourish the brain.

Optic nerve
Your eyes are actually an extension of your brain. The optic nerve and the retina grow out of the brain before birth, rather than forming separately.

Trigeminal nerve
This nerve carries signals from the face to the brain, and from the brain to chewing muscles. This is the nerve that sends pain signals when you have a toothache.

Spinal connection
The brain merges with the spinal cord at the medulla, or brain stem. The medulla triggers reflexes, such as sneezing and swallowing. It also regulates body functions such as heart rate and breathing. Just above the medulla is the pons, which serves as a link and relay station between different parts of the brain. The word "pons" means "bridge" in Latin.

▷ Cerebellum
The cerebellum sits at the lower rear of the brain, connected to the central parts by the brain stem. It has a covering of grey matter and a core of white matter. It contains an immense network of nerves that deals mostly with muscle co-ordination. As you learn to ride a bicycle or play the piano, the cerebellum is learning how to control the dozens of muscles involved. Soon you can do these movements "without thinking".

Blood supply
If the brain is deprived of oxygen for more than a few minutes it begins to die. Arteries on the brain's surface, and between the arachnoid and pia mater meninges, keep it well supplied with oxygen.

Visual centre
The visual centres at the rear of the hemispheres receive signals from the eyes. If you get knocked on the back of the head, you may "see stars" – because your visual centres are jarred and temporarily confused.

B R A I N S T E M

U P P E R R I G H T H E M I S P H E R E

W H I T E M A T T E R

L O W E R R I G H T H E M I S P H E R E

C E R E B E L L U M

Gyri (bulges) of cerebellum

Grey matter

Sulci (grooves)

The Eye

YOUR EYES ARE DELICATE organs filled with transparent jelly that measure just 25 mm (1 in) across. Deep bowls in the skull bones – the eye sockets or orbits – protect your eyeballs from damage. The exposed part of each eye is protected by the eyelids, thin folds of skin that can close rapidly.

Every time you blink, tear fluid washes over the exposed surface. A moist, clear membrane called the conjunctiva also provides lubrication. The sclera, a tough, white skin, covers most of the rest of the eyeball. Inside the sclera is a blood-rich layer called the choroid, which nourishes the other layers inside the eye. Within the choroid is the retina, where 130 million cells collect light and images to provide the picture of the world you see.

◁ Eyes left

This cross-section shows the lens on the left and the optic nerve to the right. The retina, which detects light, is microscopically thin. The vitreous body is a clear, yellowish jelly which allows light to pass through on to the retina.

Cornea

Retina

Lens

Vitreous body

Aqueous fluid

Optic nerve

THE PUPIL

The hole in the centre of the iris is called the pupil. In dim light, the pupil expands to let in as much light as possible, so you can see. In bright light, it shrinks, protecting the nerve cells at the back of the eye. Using an instrument called an ophthalmoscope, a doctor can look through the pupil and see the retina with blood vessels branching across it, as well as the optic disc, where the optic nerve leaves the eye. This is called the blind spot, because it has no nerve cells for seeing.

This is how the retina appears when viewed through an ophthalmoscope.

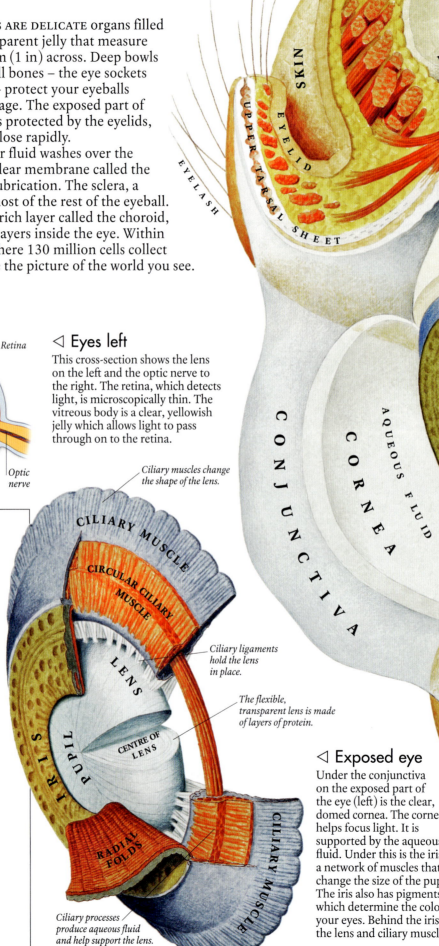

FRONTAL SINUS

FRONTAL BONE

FATTY

TEAR GLAND

SKIN

EYELID

UPPER TARSAL SHEET

EYELASH

SCLERA

CHOROID

LENS

IRIS

PUPIL

CONJUNCTIVA

CORNEA

AQUEOUS FLUID

LOWER TARSAL SHEET

EYELID

CILIARY MUSCLE

CIRCULAR CILIARY MUSCLE

LENS

IRIS

PUPIL

CENTRE OF LENS

RADIAL FOLDS

CILIARY MUSCLE

Ciliary muscles change the shape of the lens.

Ciliary ligaments hold the lens in place.

The flexible, transparent lens is made of layers of protein.

Ciliary processes produce aqueous fluid and help support the lens.

◁ Exposed eye

Under the conjunctiva on the exposed part of the eye (left) is the clear, domed cornea. The cornea helps focus light. It is supported by the aqueous fluid. Under this is the iris, a network of muscles that change the size of the pupil. The iris also has pigments which determine the colour of your eyes. Behind the iris are the lens and ciliary muscles.

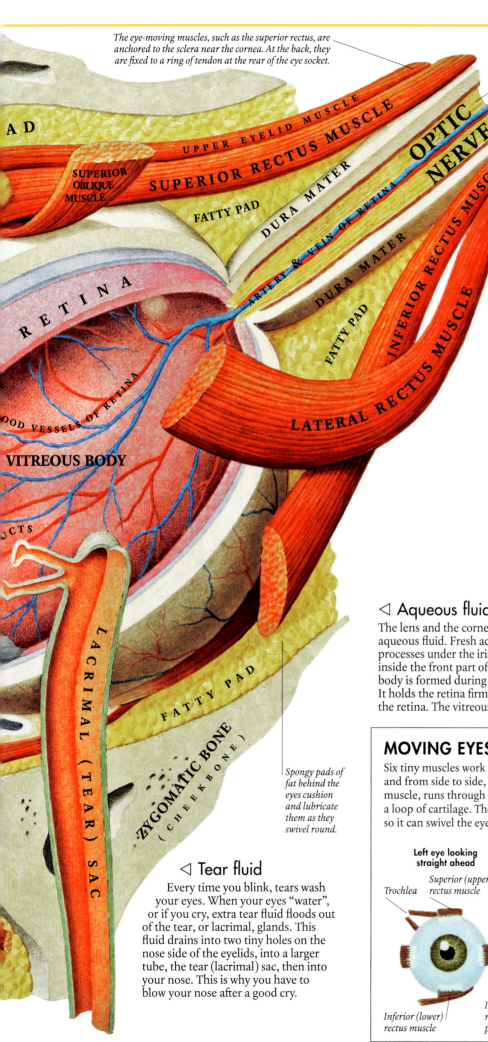

The eye-moving muscles, such as the superior rectus, are anchored to the sclera near the cornea. At the back, they are fixed to a ring of tendon at the rear of the eye socket.

UPPER EYELID MUSCLE

SUPERIOR OBLIQUE MUSCLE

SUPERIOR RECTUS MUSCLE

FATTY PAD

DURA MATER

RETINA

ARTERY & VEIN OF RETINA

OPTIC NERVE

DURA MATER

FATTY PAD

INFERIOR RECTUS MUSCLE

BLOOD VESSELS OF RETINA

VITREOUS BODY

LATERAL RECTUS MUSCLE

DUCTS

LACRIMAL (TEAR) SAC

FATTY PAD

ZYGOMATIC BONE (CHEEKBONE)

Spongy pads of fat behind the eyes cushion and lubricate them as they swivel round.

Optic nerve

When an image falls on the retina it is upside down. It is converted into nerve signals which are flashed along the optic nerve. The optic nerve passes these signals through a cross-over in the middle of the brain, known as the optic chiasma, to the visual centres at the back of the brain, where the image is turned rightside up.

RODS AND CONES

Rod cells work in dim light and see black and white. Cone cells provide colour vision but only in bright light. Each retina has about 120 million rod cells and 7 million cone cells. The rod and cone cells in the retina use a form of vitamin A to help convert light into nerve signals. The vitamin combines with proteins to make a light-sensitive chemical in the rods and colour-sensitive chemicals in the cones.

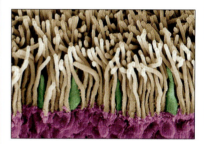

This picture shows the nerve cells in the retina, called rods and cones, magnified 1,500 times.

◁ Aqueous fluid and vitreous body

The lens and the cornea are supplied with nutrients and oxygen by the aqueous fluid. Fresh aqueous fluid is constantly produced by the ciliary processes under the iris. This maintains the correct level of pressure inside the front part of the eye, and helps hold its shape. The vitreous body is formed during development in the uterus, and is not replaced. It holds the retina firmly against the choroid, and transmits light on to the retina. The vitreous body also gives the eyeball its spherical shape.

MOVING EYES

Six tiny muscles work as a team to make each eye look up or down and from side to side, at any angle. One of these, the superior oblique muscle, runs through a tiny "pulley" known as a trochlea, formed by a loop of cartilage. The trochlea alters the muscle's direction of pull, so it can swivel the eye to look from side to side.

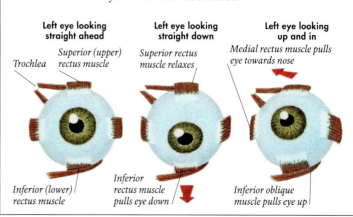

Left eye looking straight ahead

Trochlea

Superior (upper) rectus muscle

Inferior (lower) rectus muscle

Left eye looking straight down

Superior rectus muscle relaxes

Inferior rectus muscle pulls eye down

Left eye looking up and in

Medial rectus muscle pulls eye towards nose

Inferior oblique muscle pulls eye up

◁ Tear fluid

Every time you blink, tears wash your eyes. When your eyes "water", or if you cry, extra tear fluid floods out of the tear, or lacrimal, glands. This fluid drains into two tiny holes on the nose side of the eyelids, into a larger tube, the tear (lacrimal) sac, then into your nose. This is why you have to blow your nose after a good cry.

Inside the Ear

FROM QUIET WHISPERS TO NOISY CRASHES your ears pick up a huge array of sounds. Your ears and brain together provide your sense of hearing. Ears convert sound waves into electrical nerve impulses, which speed to the brain to be deciphered. They also help you to keep your balance by telling your brain which way up you are.

What we usually call our "ear", the curvy flap of skin on the side of the head, catches sound waves and funnels them into a tube called the outer ear canal. The canal leads deep inside the skull, where the sound waves reach the parts of the ear that actually do the hearing. The conversion of sounds into electrical impulses takes place in a fluid-filled spiral structure, the cochlea, which could sit on your thumbnail like a small snail. Attached to the cochlea are three C-shaped canals, which help you to keep your balance.

▷ Peer through an ear

There are three main areas of the ear, the outer, middle, and inner ears. The outer ear is made up of the curved flap of skin and cartilage on the outside of your head and the ear canal. The canal stretches to the middle ear, a tiny chamber containing the three auditory (hearing) bones. Beyond these bones are the structures of the inner ear, the cochlea and semicircular canals.

▷ Hearing aid

Your ear flap is also called the auricle or pinna. Its curvy rims, the outer and inner helixes, help to collect sounds. In the past, people who had trouble hearing sometimes used an ear trumpet. This device collects more sound waves and concentrates them for the ear canal.

Ear trumpet

The ear canal
The hole in the middle of your outer ear is the entrance to the outer ear canal, also known as the external auditory canal. Sound waves travel along this 2.5 cm (1 in) passageway before reaching the eardrum.

Wax in your ears
Ear wax, or cerumen, is made by glands in the skin that line the ear canal. It works with the hairs in the canal to trap dirt and dust, before they can reach your delicate eardrum.

AIR TO SOLID TO LIQUID

Sound reaches your ear as waves of vibrating air molecules. When these vibrations strike and rattle the eardrum, its movements make the three linked ear bones in the middle ear vibrate. The third bone passes the vibrations to the fluid inside the cochlea, in the inner ear. This process is called air conduction.

Your own voice reaches the inner ear through your skull bones, a process known as bone conduction. When you listen to your voice on a tape recording, it might sound strange. That is because the sound is reaching your ears through air instead of partly through your skull bones, as you usually hear it.

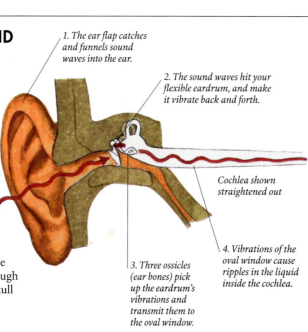

1. The ear flap catches and funnels sound waves into the ear.

2. The sound waves hit your flexible eardrum, and make it vibrate back and forth.

Cochlea shown straightened out

3. Three ossicles (ear bones) pick up the eardrum's vibrations and transmit them to the oval window.

4. Vibrations of the oval window cause ripples in the liquid inside the cochlea.

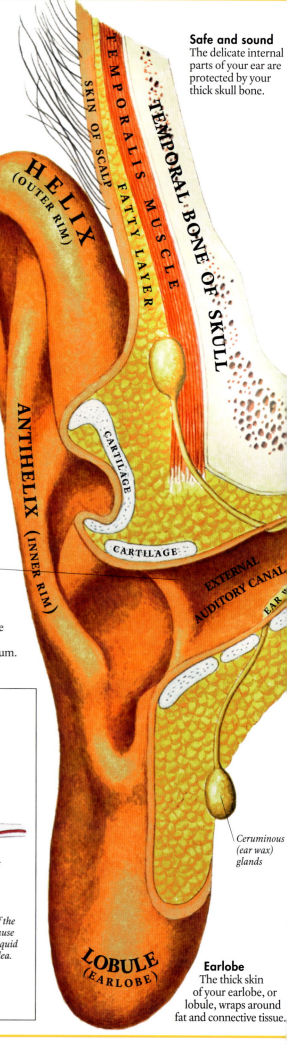

Safe and sound
The delicate internal parts of your ear are protected by your thick skull bone.

HELIX (OUTER RIM)
SKIN OF SCALP
TEMPORALIS MUSCLE
TEMPORAL BONE OF SKULL
FATTY LAYER
CARTILAGE
ANTIHELIX (INNER RIM)
CARTILAGE
EXTERNAL AUDITORY CANAL
LOBULE (EARLOBE)

Ceruminous (ear wax) glands

Earlobe
The thick skin of your earlobe, or lobule, wraps around fat and connective tissue.

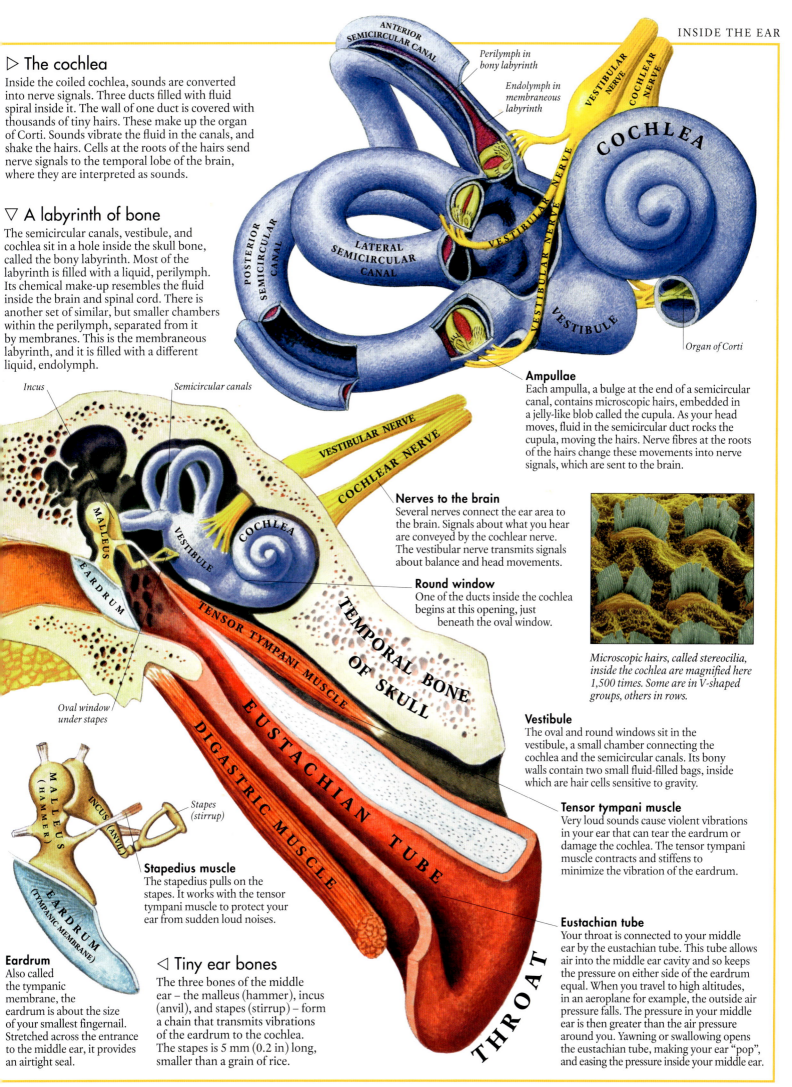

▷ The cochlea

Inside the coiled cochlea, sounds are converted into nerve signals. Three ducts filled with fluid spiral inside it. The wall of one duct is covered with thousands of tiny hairs. These make up the organ of Corti. Sounds vibrate the fluid in the canals, and shake the hairs. Cells at the roots of the hairs send nerve signals to the temporal lobe of the brain, where they are interpreted as sounds.

▽ A labyrinth of bone

The semicircular canals, vestibule, and cochlea sit in a hole inside the skull bone, called the bony labyrinth. Most of the labyrinth is filled with a liquid, perilymph. Its chemical make-up resembles the fluid inside the brain and spinal cord. There is another set of similar, but smaller chambers within the perilymph, separated from it by membranes. This is the membraneous labyrinth, and it is filled with a different liquid, endolymph.

ANTERIOR SEMICIRCULAR CANAL

Perilymph in bony labyrinth

Endolymph in membraneous labyrinth

VESTIBULAR NERVE

COCHLEAR NERVE

COCHLEA

POSTERIOR SEMICIRCULAR CANAL

LATERAL SEMICIRCULAR CANAL

VESTIBULAR NERVE

VESTIBULE

Organ of Corti

Ampullae

Each ampulla, a bulge at the end of a semicircular canal, contains microscopic hairs, embedded in a jelly-like blob called the cupula. As your head moves, fluid in the semicircular duct rocks the cupula, moving the hairs. Nerve fibres at the roots of the hairs change these movements into nerve signals, which are sent to the brain.

Incus

Semicircular canals

VESTIBULAR NERVE

COCHLEAR NERVE

MALLEUS

VESTIBULE

COCHLEA

EARDRUM

Nerves to the brain

Several nerves connect the ear area to the brain. Signals about what you hear are conveyed by the cochlear nerve. The vestibular nerve transmits signals about balance and head movements.

Round window

One of the ducts inside the cochlea begins at this opening, just beneath the oval window.

TENSOR TYMPANI MUSCLE

TEMPORAL BONE OF SKULL

Microscopic hairs, called stereocilia, inside the cochlea are magnified here 1,500 times. Some are in V-shaped groups, others in rows.

Vestibule

The oval and round windows sit in the vestibule, a small chamber connecting the cochlea and the semicircular canals. Its bony walls contain two small fluid-filled bags, inside which are hair cells sensitive to gravity.

Oval window under stapes

EUSTACHIAN TUBE

DIGASTRIC MUSCLE

Tensor tympani muscle

Very loud sounds cause violent vibrations in your ear that can tear the eardrum or damage the cochlea. The tensor tympani muscle contracts and stiffens to minimize the vibration of the eardrum.

MALLEUS (HAMMER)

INCUS (ANVIL)

Stapes (stirrup)

Stapedius muscle

The stapedius pulls on the stapes. It works with the tensor tympani muscle to protect your ear from sudden loud noises.

EARDRUM (TYMPANIC MEMBRANE)

Eardrum

Also called the tympanic membrane, the eardrum is about the size of your smallest fingernail. Stretched across the entrance to the middle ear, it provides an airtight seal.

◁ Tiny ear bones

The three bones of the middle ear – the malleus (hammer), incus (anvil), and stapes (stirrup) – form a chain that transmits vibrations of the eardrum to the cochlea. The stapes is 5 mm (0.2 in) long, smaller than a grain of rice.

THROAT

Eustachian tube

Your throat is connected to your middle ear by the eustachian tube. This tube allows air into the middle ear cavity and so keeps the pressure on either side of the eardrum equal. When you travel to high altitudes, in an aeroplane for example, the outside air pressure falls. The pressure in your middle ear is then greater than the air pressure around you. Yawning or swallowing opens the eustachian tube, making your ear "pop", and easing the pressure inside your middle ear.

The Neck

IMAGINE HAVING a permanently stiff neck! You would have to move your whole body just to turn your head. The neck provides a strong and flexible stalk for the head, allowing it to tilt and twist. To hold your head steady and move your neck, you rely on sets of muscles that link the neck with the spine, ribs, sternum, and shoulder bones.

Three vital sets of pipework pass through the neck. One is the tunnel down the middle of the spine, which houses the spinal cord, the main nerve in the body. The second is the trachea, or windpipe, which conveys air to and from the lungs. The third is the oesophagus, or gullet. When you swallow, food from the mouth passes down the oesophagus, through the chest, to the stomach. All three sets of pipework are stretchy and pliable, bending easily as you move. There are also blood vessels, nerves, and lymph vessels in the neck.

▽ Foramen magnum

Looking at the skull from below, you can see a large hole called the foramen magnum. This is where the spinal cord passes through to join the base of the brain.

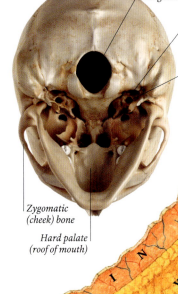

Foramen magnum

Hole for jugular vein

Hole for carotid artery

Zygomatic (cheek) bone

Hard palate (roof of mouth)

SEMISPINALIS CAPITIS MUSCLE

SPLENIUS CAPITIS MUSCLE

LEVATOR SCAPULAE MUSCLE

SKIN

FAT LAYER

BACK MUSCLE

DELTOID MUSCLE

SCAPULA

BACK MUSCLE

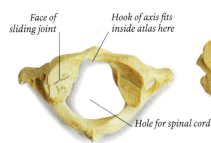

Face of sliding joint

Hook of axis fits inside atlas here

Peglike hook (dens)

Hole for spinal cord

△ Atlas

This bone sits at the top of the spinal column. It is more ring-shaped than the others. Sliding joints on each side of it allow the head to nod up and down.

△ Axis

Directly under the atlas is the axis. It has a peglike hook (the dens) that fits into a notch in the atlas. This peg- and-ring system allows the head to swivel from side to side.

△ Levator scapulae

This long muscle is fixed to the atlas, axis, and neck backbones (vertebrae) above, and to the shoulder blade (scapula) below. When you carry a weight on your shoulder, the levator scapulae ("shoulder blade lifter") tenses up, becoming hard and stiff.

NECK MUSCLES

Each pair of neck muscles pulls the skull in a different way. When you look up, the vertical muscles at the back of your neck contract. When you look down, the muscles at the front contract. Other neck muscles run diagonally around the neck. They pull the skull around, letting you twist your head. At the same time, opposing muscles tense so that the face is kept steady and looking horizontally, rather than being tilted down.

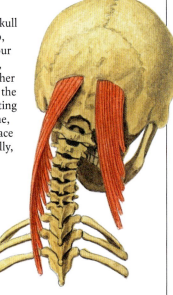

The two splenius capitis muscles, shown here, are one of the sets of muscles that join the base of the skull and the spine. They rotate the face and tilt the head from side to side.

SPINAL NERVE ROOTS

Spinal nerves

Extending from the spinal cord are 31 pairs of spinal nerves. They emerge from a bulge at the top of the spinal cord called the cervical enlargement.

SKULL

TEMPORALIS MUSCLE

EYE-CLOSING MUSCLE

EYE

FAT LAYER

ZYGOMATIC BONE

ZYGOMATIC MUSCLE

TEMPORALIS VEIN

JAWBONE

STERNOCLEIDOMASTOID MUSCLE

MASSETER MUSCLE

JUGULAR VEIN

FACIAL ARTERY

SPINE

LONGISSIMUS CAPITIS MUSCLE

ILIOCOSTALIS CERVICIS MUSCLE

FAT LAYER

No nose bones
The front part of the nose does not contain bones. It is supported by a framework of nine gristle-like cartilages, joined to each other and to the skull bones. One of these, the septal cartilage, divides the nostrils.

Cheek muscle
"Zygomatic" is another word for "cheek". To smile, you use your zygomaticus major muscle, which connects the corner of your mouth to your zygomatic bone. With other muscles' help, it pulls the lips wider and upward, into a grin.

Trapezius muscle
The trapezius – one of the biggest shoulder muscles – is flat and triangular. It helps to turn and tilt the head, raise and twist the arm, and shrug or steady the shoulder.

TRAPEZIUS MUSCLE

FAT

SCAPULA

INFRASPINATUS MUSCLE

TERES MINOR MUSCLE

Transverse process

△ Stand up straight and tall
The longissimus capitis is one of a series of muscles that are anchored at various points from the skull and neck vertebrae down to the lower back. This muscle group, which also includes the longissimus cervicis, is collectively known as the erector spinae muscles because they keep the spine erect – that is, keep you standing up straight!

Spinal bones
Most of the spinal vertebrae have two flanges or wings, one on either side, called transverse processes. These are anchor points for the spinal muscles.

OESOPHAGUS AND TRACHEA
Air can enter the body through both the nose and the mouth. Food comes through the mouth. The passageways from the nose and mouth join together behind the tongue to form the throat, or pharynx. Farther down they split again, into the oesophagus, for food, and the trachea, for air, with the larynx at its top.

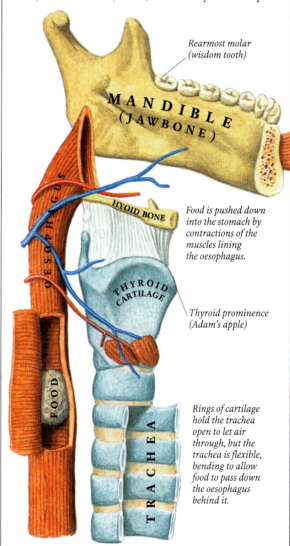

Rearmost molar (wisdom tooth)

MANDIBLE (JAWBONE)

OESOPHAGUS

HYOID BONE

Food is pushed down into the stomach by contractions of the muscles lining the oesophagus.

THYROID CARTILAGE

Thyroid prominence (Adam's apple)

FOOD

Rings of cartilage hold the trachea open to let air through, but the trachea is flexible, bending to allow food to pass down the oesophagus behind it.

TRACHEA

DOWN THE GULLET
The oesophagus, or gullet, is normally squashed flat by internal body pressure. When you swallow, muscles arranged in circles within the wall of the oesophagus contract in sequence. The travelling waves of contraction push food or drink downward.

When food enters the pharynx, a flap of muscle swings across the entrance of the trachea to prevent the food from entering it. If something you eat "goes down the wrong tube", coughing forces air up the trachea, which loosens and ejects the food.

The Upper Torso

UNDERNEATH THE SKIN of your chest lie the heart and lungs, working together to supply your entire body with fresh, oxygen-rich blood. A network of broad, flat muscles covers the moveable, bony cage formed by the ribs. The rib cage protects the soft, spongy lungs and pumping heart. Two vertical bony structures, the sternum (breastbone) at the front and the spinal column at the rear, act as girders to provide strength for the rib cage. Joints between the ribs, sternum, and spinal column allow the cage to change shape, so that the lungs can expand and contract as you breathe.

The cone-shaped area of your chest is also known as the thorax, and the belly below it is termed the abdomen. The two sections are divided just below your lungs by the diaphragm, a powerful sheet of muscle shaped like an upside-down bowl.

▷ Mammary gland

Everyone has mammary glands, but they are not highly developed in children or men. A woman's glands become active after childbirth, making milk to feed the baby. Each mammary gland has about 20 rounded lobes, held together by loose connective tissue. Each lobe contains numerous milk-making lobules. Narrow tubes, called lactiferous ducts, carry the milk into tiny openings in the nipple.

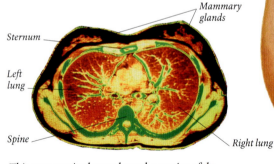

Mammary glands
Sternum
Left lung
Spine
Right lung

This computerized scan through a section of the chest shows the lungs, with the spine behind and the sternum and mammary glands in front.

▷ Inside the chest

Your chest must be strong enough to protect the delicate organs within, yet flexible enough to move as you breathe. The sheet-like layers of muscle anchored to your clavicle, or collar bone, contract to lift and expand your chest as you breathe in and relax as you breathe out.

Alveoli

The milk made in the grape-like clusters of the alveoli glands is rich in nutrients and also protects the baby against some diseases.

Lactiferous duct

When a baby sucks the nipple, milk enters the lactiferous ducts and travels to tiny openings on its surface.

LYMPH DUCTS

PECTORALIS MAJOR MUSCLE

SKIN

BREAST

AREOLA

NIPPLE

ADIPOSE (FATTY) TISSUE

SKIN

SKIN

FATTY LAYER

CLAVICLE

DELTOID MUSCLE

AXILLARY VEIN

PECTORALIS MAJOR MUSCLE

SKIN

MUSCLE

BREAST

FATTY LAYER

SKIN

SEVENTH RIB

EIGHTH RIB

NINTH RIB

TENTH RIB

A FLEXIBLE CAGE

Your rib cage is made of 12 pairs of springy C-shaped bones. The first pair is short; the pairs below lengthen, then shorten again, to form the cage. Joined to vertebrae at the back, each rib arches around and down towards the sternum at the front. But the rib bones themselves do not reach the sternum. Most are connected to the sternum by costal cartilage – tough, rubbery bars that give the entire rib cage extra flexibility.

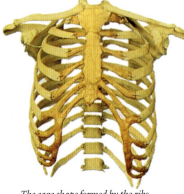

The cage shape formed by the ribs protects delicate vital organs inside.

This suit of armour is like an extra rib cage, worn on the outside.

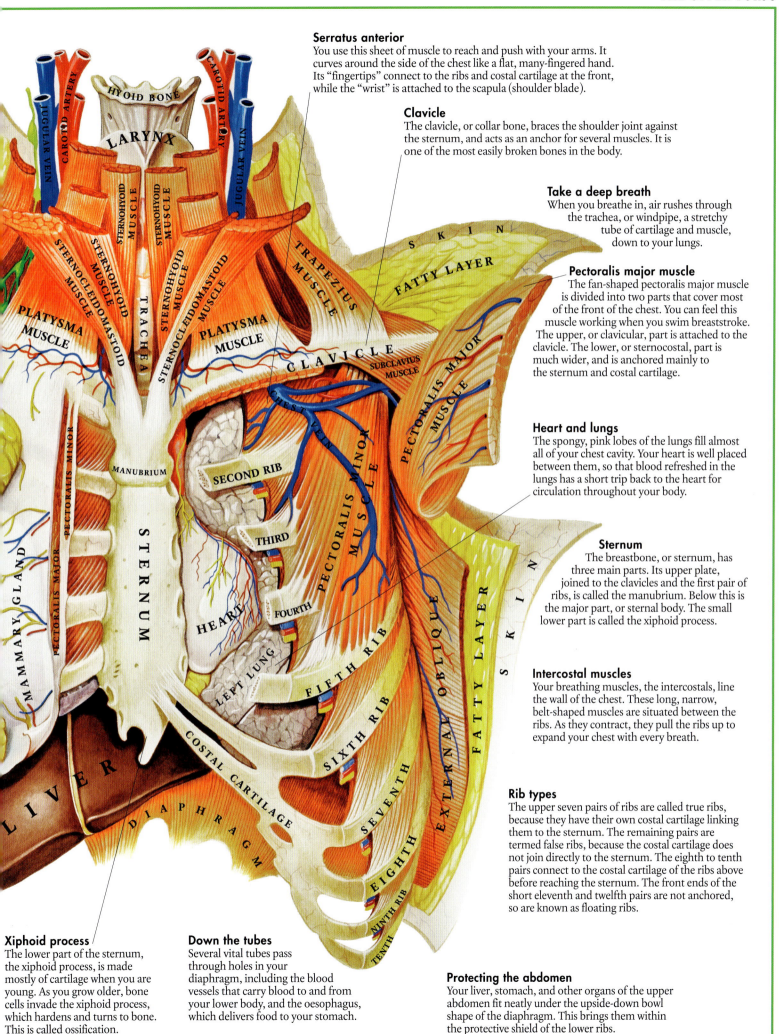

Serratus anterior
You use this sheet of muscle to reach and push with your arms. It curves around the side of the chest like a flat, many-fingered hand. Its "fingertips" connect to the ribs and costal cartilage at the front, while the "wrist" is attached to the scapula (shoulder blade).

Clavicle
The clavicle, or collar bone, braces the shoulder joint against the sternum, and acts as an anchor for several muscles. It is one of the most easily broken bones in the body.

Take a deep breath
When you breathe in, air rushes through the trachea, or windpipe, a stretchy tube of cartilage and muscle, down to your lungs.

Pectoralis major muscle
The fan-shaped pectoralis major muscle is divided into two parts that cover most of the front of the chest. You can feel this muscle working when you swim breaststroke. The upper, or clavicular, part is attached to the clavicle. The lower, or sternocostal, part is much wider, and is anchored mainly to the sternum and costal cartilage.

Heart and lungs
The spongy, pink lobes of the lungs fill almost all of your chest cavity. Your heart is well placed between them, so that blood refreshed in the lungs has a short trip back to the heart for circulation throughout your body.

Sternum
The breastbone, or sternum, has three main parts. Its upper plate, joined to the clavicles and the first pair of ribs, is called the manubrium. Below this is the major part, or sternal body. The small lower part is called the xiphoid process.

Intercostal muscles
Your breathing muscles, the intercostals, line the wall of the chest. These long, narrow, belt-shaped muscles are situated between the ribs. As they contract, they pull the ribs up to expand your chest with every breath.

Rib types
The upper seven pairs of ribs are called true ribs, because they have their own costal cartilage linking them to the sternum. The remaining pairs are termed false ribs, because the costal cartilage does not join directly to the sternum. The eighth to tenth pairs connect to the costal cartilage of the ribs above before reaching the sternum. The front ends of the short eleventh and twelfth pairs are not anchored, so are known as floating ribs.

Xiphoid process
The lower part of the sternum, the xiphoid process, is made mostly of cartilage when you are young. As you grow older, bone cells invade the xiphoid process, which hardens and turns to bone. This is called ossification.

Down the tubes
Several vital tubes pass through holes in your diaphragm, including the blood vessels that carry blood to and from your lower body, and the oesophagus, which delivers food to your stomach.

Protecting the abdomen
Your liver, stomach, and other organs of the upper abdomen fit neatly under the upside-down bowl shape of the diaphragm. This brings them within the protective shield of the lower ribs.

The Upper Torso – Back

MANY OF THE MUSCLES that give strength to the back of your upper torso are near the surface of the skin. But underneath these muscles lies the real secret of your upper body's support: the strong, T-shaped structure formed by the spinal column, the rib cage, and the pair of scapulae (shoulder blades) linked to the clavicles (collar bones).

The ribs wrap up and around the front of your chest, and meet the spinal column at the rear. The ends of each pair of ribs form joints with the vertebrae, or spinal bones. Together with the scapulae, these bones form a support system strong enough to bear the heavy weight of the muscles, bones, and organs of the head and chest, yet flexible enough to let you stretch up, reach across your body, bend double, and twist your body from side to side. A deep layer of muscles, some interwoven with the ribs and others linked to bony knobs on the spine, provides added flexibility and stability.

Deltoid

Trapezius

Triceps

Latissimus dorsi

The outlines of some of this bodybuilder's superficial muscles – those just under the skin – are clearly visible. The muscles beneath these are called deep muscles.

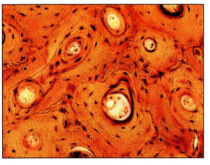

This is a micrograph of compact bone, like that forming the outer layer of a vertebra. Inside each round column of bone cells is a space called the haversian canal, containing nerves and blood and lymph vessels.

▽ Inside a vertebra

The typical vertebra has an outer shell of hard, dense bone, called compact bone. Within is the honeycomb of cancellous bone, which contains red bone marrow. All of your red blood cells and many of your white cells are made inside this soft, fatty tissue. A hormone signal starts the production of new cells in the marrow. There, they mature before they are released into circulation.

▷ The backbone of your body

The moveable joints between the rear end of a rib and the side of a vertebra are called costovertebral joints. In fact, each of these is a double joint. The end of the rib fits snugly into a shallow socket on the body (main part) of the vertebra, while a rear-facing part of the rib, called the costal neck and tubercle (bump), fits into the transverse process or "side wing" of the vertebra.

▷ Inside a rib

Under its thin shell of compact bone, the rib is richly supplied with red bone marrow. Not all bones contain red marrow – it is found only within the ribs, vertebrae, sternum (breast bone), clavicles, scapulae, pelvis (hip bones), and skull bones. Other bones can contain yellow marrow, which is mainly made of fat and other tissues.

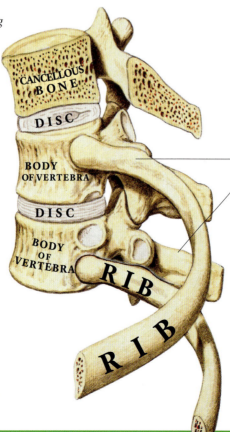

CANCELLOUS BONE

DISC

BODY OF VERTEBRA

DISC

BODY OF VERTEBRA

RIB

RIB

RIB

Costal neck and tubercle of rib

Transverse process

Latissimus dorsi

The base of this long triangle of muscle is attached to the chest vertebrae, the lower ribs, and even to the hip bones far below. The muscle fibres at its tip meet under the scapula and join to the upper arm bone, the humerus, in the shoulder. You use the latissimus dorsi as you swing your arms back when jogging, or reach up to grab something above your head.

TRAPEZIUS

DELTOID

TERES MAJOR

SERRATUS ANTERIOR

SKIN

LATISSIMUS DORSI

FATTY LAYER

▷ Rear of the rib cage

A back view of the thorax (chest) shows one pair of ribs for each thoracic vertebra. Each scapula joins to the clavicle and to the upper arm bone in the shoulder, but there are no direct joints between the scapula and ribs. This means the scapula can slide around the back of the rib cage, contributing to the shoulder's great flexibility.

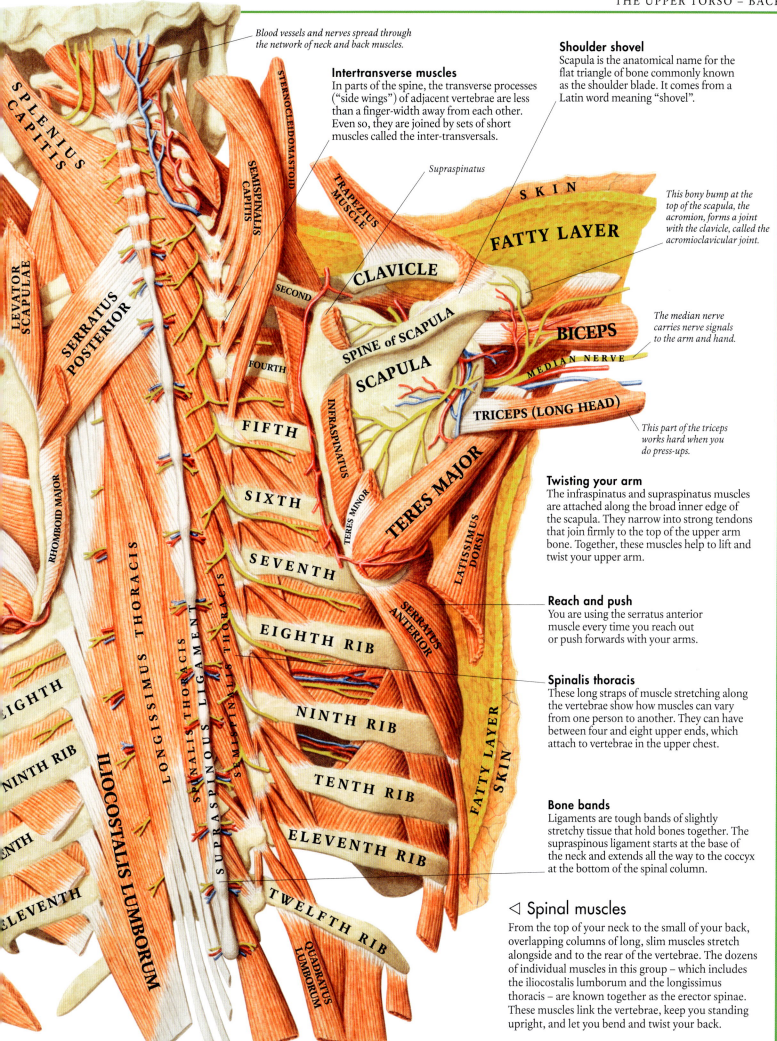

SPLENIUS CAPITIS

Blood vessels and nerves spread through the network of neck and back muscles.

STERNOCLEIDOMASTOID

SEMISPINALIS CAPITIS

LEVATOR SCAPULAE

SERRATUS POSTERIOR

TRAPEZIUS MUSCLE

RHOMBOID MAJOR

SECOND

FOURTH

Intertransverse muscles

In parts of the spine, the transverse processes ("side wings") of adjacent vertebrae are less than a finger-width away from each other. Even so, they are joined by sets of short muscles called the inter-transversals.

Supraspinatus

Shoulder shovel

Scapula is the anatomical name for the flat triangle of bone commonly known as the shoulder blade. It comes from a Latin word meaning "shovel".

SKIN

FATTY LAYER

This bony bump at the top of the scapula, the acromion, forms a joint with the clavicle, called the acromioclavicular joint.

CLAVICLE

SPINE of SCAPULA

SCAPULA

INFRASPINATUS

TERES MINOR

BICEPS

MEDIAN NERVE

The median nerve carries nerve signals to the arm and hand.

TRICEPS (LONG HEAD)

This part of the triceps works hard when you do press-ups.

TERES MAJOR

LATISSIMUS DORSI

SERRATUS ANTERIOR

FIFTH

SIXTH

SEVENTH

EIGHTH RIB

NINTH RIB

TENTH RIB

ELEVENTH RIB

TWELFTH RIB

ILIOCOSTALIS LUMBORUM

LONGISSIMUS THORACIS

SPINALIS THORACIS

SUPRASPINOUS LIGAMENT

SEMISPINALIS THORACIS

QUADRATUS LUMBORUM

EIGHTH

NINTH RIB

TENTH

ELEVENTH

FATTY LAYER SKIN

Twisting your arm

The infraspinatus and supraspinatus muscles are attached along the broad inner edge of the scapula. They narrow into strong tendons that join firmly to the top of the upper arm bone. Together, these muscles help to lift and twist your upper arm.

Reach and push

You are using the serratus anterior muscle every time you reach out or push forwards with your arms.

Spinalis thoracis

These long straps of muscle stretching along the vertebrae show how muscles can vary from one person to another. They can have between four and eight upper ends, which attach to vertebrae in the upper chest.

Bone bands

Ligaments are tough bands of slightly stretchy tissue that hold bones together. The supraspinous ligament starts at the base of the neck and extends all the way to the coccyx at the bottom of the spinal column.

◁ Spinal muscles

From the top of your neck to the small of your back, overlapping columns of long, slim muscles stretch alongside and to the rear of the vertebrae. The dozens of individual muscles in this group – which includes the iliocostalis lumborum and the longissimus thoracis – are known together as the erector spinae. These muscles link the vertebrae, keep you standing upright, and let you bend and twist your back.

The Lungs

THE LUNGS TAKE IN air and allow oxygen to pass into the blood which flows through them, while removing waste carbon dioxide from the blood, which is then carried away in the air you breathe out. The main bulk of the lungs is composed of tiny, branching air tubes called bronchioles, which end in microscopic "air bubbles" called alveoli. Each lung has 300–350 million alveoli. Gas exchange – oxygen for carbon dioxide – takes place inside these alveoli, where air and blood are separated by a film of moisture and membranes only 0.0002 mm (0.000008 in) thick.

▷ The lobes of your lungs

You have two lungs, but they are not exactly the same. The right one has three lobes, or sections. The left one has two lobes and a hollowed-out area to make room for the heart. The upper tips (or apices) of the lungs reach above the inner ends of the clavicles (collar bones), almost into the neck. The lungs rest on the dome-shaped main breathing muscle, the diaphragm.

Pleural membranes

The lungs are enveloped within two smooth layers of membrane, the pleurae. The inner, or pulmonary, layer follows every contour of the lungs, wrapping around the lobes and the main airways. It folds back and over on itself, to form the outer, or parietal, layer.

Between the two pleural layers is a tiny space containing slippery pleural fluid. The fluid works with the pleurae to lubricate the lungs as you breathe.

BREATHING IN AND OUT

When you breathe in (inhale), your diaphragm tenses and its dome shape flattens, while your chest muscles pull the ribs up and out. Your chest cavity and lungs expand, as air rushes down the trachea and into the spongy lungs. When you breathe out (exhale), these two sets of muscles relax. Your ribs fall back, while the pressure from the abdominal contents below pushes your diaphragm back up. Air is forced out, and the lungs spring back to their starting size. When you are relaxed you breathe about 15 times each minute.

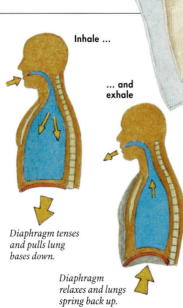

Inhale ...

... and exhale

Diaphragm tenses and pulls lung bases down.

Diaphragm relaxes and lungs spring back up.

△ **Lung layers**

This view shows the layers in the lungs, from the intercostal muscles between ribs on the outside to the lung tissue on the inside.

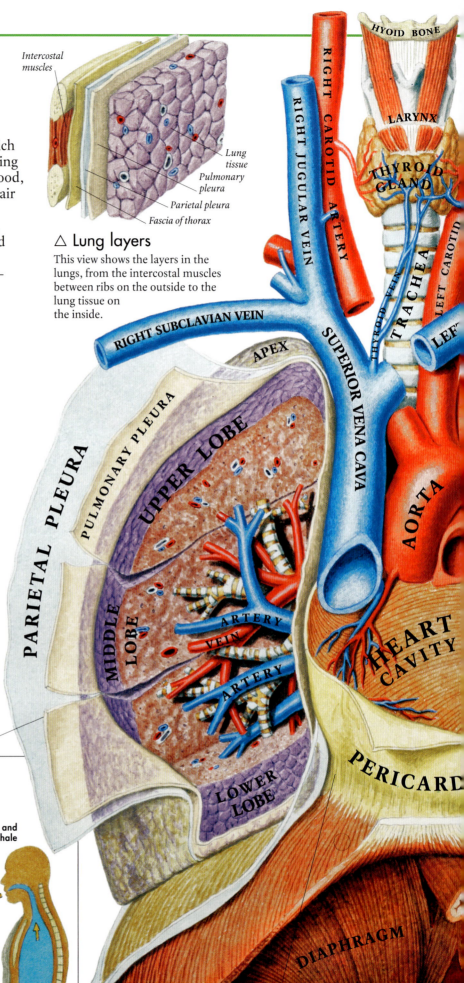

Intercostal muscles

Lung tissue
Pulmonary pleura
Parietal pleura
Fascia of thorax

HYOID BONE

RIGHT CAROTID ARTERY

RIGHT JUGULAR VEIN

LARYNX

THYROID GLAND

THYROID VEIN

LEFT CAROTID

TRACHEA

LEFT

RIGHT SUBCLAVIAN VEIN

SUPERIOR VENA CAVA

APEX

PARIETAL PLEURA

PULMONARY PLEURA

UPPER LOBE

AORTA

MIDDLE LOBE

ARTERY

VEIN

ARTERY

HEART CAVITY

LOWER LOBE

PERICARD

DIAPHRAGM

ABDOMINAL AORTA

Pericardium
This tough membrane encloses the heart and holds it in place.

Air conditioning

Before air travels to the lungs, it is "treated" inside your nose. Nasal hairs and sticky mucus trap dust and other particles. Cold air is warmed and dry air is moistened to match the conditions within the lungs.

Trachea

Beneath the larynx is the trachea, the tube that carries air into the lungs. About 20 C-shaped hoops of cartilage help keep it open at all times.

Hairy lining

Thousands of tiny hairs called cilia line the walls of your airways. They move in waves to push mucus and specks of dust up out of the lungs.

Venule

Arteriole

SUBCLAVIAN VEIN

Mucous membrane

Elastic fibres

CAPILLARY NETWORK

Cartilage

Smooth muscles

ALVEOLI

UPPER LOBE

LEFT PULMONARY VEIN

ARTERY

VEIN

BRONCHUS

BRONCHIOLE

LOWER LOBE

SEPTUM

Lung capacity

You normally breathe in about half a litre (about a pint) of air around 15 times a minute. But when you exercise, you breathe more deeply and twice as often, so you take in nearly five times as much air.

Three trees in one

The structures inside your lungs resemble three upside down trees. One "tree" is the airway: the trachea branches into smaller bronchi, which divide into bronchioles. Another is the pulmonary artery bringing stale blood (shown in blue) from the heart, which divides to form the pulmonary arterioles and finally the alveolar capillaries. The third is the network of pulmonary veins and venules taking refreshed blood (in red) back to the heart.

▷ Inside an alveolus

The lining of the alveoli is made of a single layer of thin, flat, curved cells. The walls of the capillaries around the alveoli have a similar construction. As a result, air inside the alveolus is extremely close to the blood cells in the capillary. This means the oxygen in your lungs only has a tiny distance to travel in order to enter the bloodstream.

Diaphragm

Signals from your brain make your diaphragm contract without you telling it to.

DIAPHRAGM

Hiccups

The reflex action that causes hiccups happens in two stages. First, the diaphragm contracts sharply when its nerves become irritated. Then, as soon as you breathe in, a flap of skin over your oesophagus snaps shut, making a clicking sound.

▽ Alveoli

The smallest lung airways, the terminal bronchioles, look like tiny twigs of a vine bearing bunches of grapes. The "grapes" are groups of air sacs called alveoli. Around 700 million of them are clustered inside your lungs. Pulmonary arterioles branch and divide to form the smallest blood vessels, the capillaries, which wrap around each alveolus.

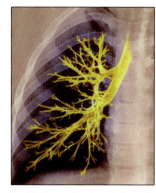

In this X-ray, the airways branch out like a tree in the chest cavity, encased by the ribs.

Mucous membranes

Lining the respiratory system are mucous membranes, which make sticky mucus to keep the structures within moist. Mucus also helps to trap dust.

A quick exchange

The alveoli increase the surface area of the lung, so that exchanging oxygen and carbon dioxide is quick and efficient. If all the alveoli from both lungs were spread flat, they would cover an area nearly the size of a tennis court.

Yawning

When you open your mouth wide in a deep yawn, air rushes into your lungs, expanding the alveoli. No one is quite sure what triggers the yawning reflex, but it usually follows a period of shallow breathing, when you are tired or stressed.

Macrophages

These tiny scavenger cells inside the alveolus "eat" and destroy any dust or particles of bacteria that manage to reach your lungs.

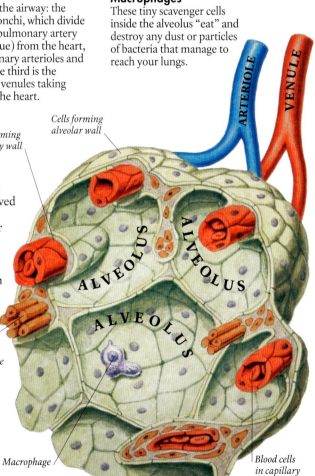

Cells forming capillary wall

Cells forming alveolar wall

ARTERIOLE

VENULE

ALVEOLUS

ALVEOLUS

ALVEOLUS

Fibrous supporting tissue

Macrophage

Blood cells in capillary

The Heart and Main Blood Vessels

NESTLING BETWEEN THE lower parts of your lungs is the heart, a fist-sized bag of muscle. Consisting of two pumps, it contracts and squeezes out its contents more than once a second. It works every minute of every day, pumping blood around your body's 150,000-km (90,000-mile) network of blood vessels.

A wall called the septum separates the two pumps in the heart. The pump on the left side (shown on the right in the pictures) squirts bright red oxygen-rich blood into the main artery, the aorta. This leads into a network of blood vessels that reach every part of your body. The blood passes oxygen on to the body tissues, and returns, now reddish-purple in colour, to the right side of the heart. From here it is pumped out to the lungs, where it is refreshed with extra oxygen before returning to the left pump to complete its circuit again.

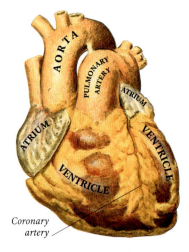

▷ Coronary vessels

Like all body organs, your heart needs a supply of blood to bring it oxygen. The muscle that makes up the wall of the heart, the myocardium, receives oxygen-rich blood from a system of small arteries that branch from the aorta. These are called coronary arteries. They snake over the heart's surface, dividing and sending tiny branches into the heart.

Coronary artery

HOW VALVES WORK

A series of valves ensure that blood only flows one way. At each exit leading from the heart into the main arteries is a valve made of three flexible flaps. When the heart pumps, these flaps flatten against the artery wall, allowing blood to surge past in the correct direction. When the heart refills, they balloon out and seal along their edges to stop the blood flowing backwards.

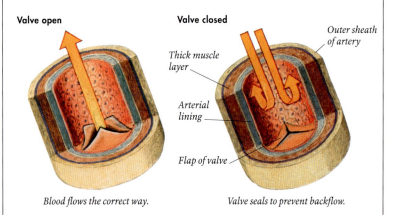

Valve open

Valve closed

Outer sheath of artery

Thick muscle layer

Arterial lining

Flap of valve

Blood flows the correct way.

Valve seals to prevent backflow.

Superior vena cava
This vein returns used blood from the head, arms, and upper body to the heart. The inferior vena cava returns blood from the lower body and legs.

Right coronary artery
No wider than a drinking straw, this loops around the heart's lower side and to the rear. It supplies blood to the lower pumping chamber, the ventricle.

Pulmonary arteries
The two pulmonary arteries carry dark, low-oxygen blood from the heart to the lungs for replenishment.

Flexible muscle and tissue withstand blood pressure.

Atrium and ventricle
The heart consists of four chambers. The right atrium and right ventricle form the right pump, and the left atrium and left ventricle form the left pump. The atriums receive blood from the veins and pass it into the ventricles.

Tricuspid valve
A three-pointed valve between the right atrium and ventricle prevents blood surging back into the atrium as the ventricle contracts powerfully. Its thin anchoring tendons prevent it turning inside out.

Interventricular septum
This dividing wall is slightly off-centre. That means the left ventricle – the one with the greater job of pumping blood all around the body – is larger, and also more muscular, than the right ventricle.

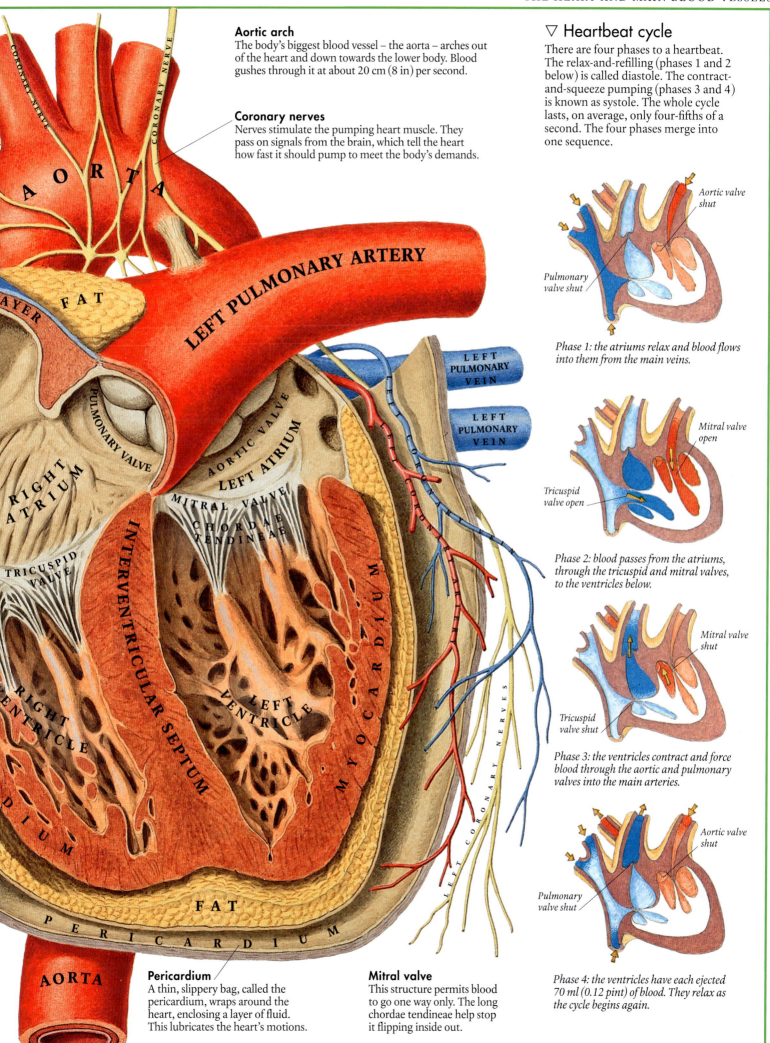

Aortic arch
The body's biggest blood vessel – the aorta – arches out of the heart and down towards the lower body. Blood gushes through it at about 20 cm (8 in) per second.

Coronary nerves
Nerves stimulate the pumping heart muscle. They pass on signals from the brain, which tell the heart how fast it should pump to meet the body's demands.

LEFT PULMONARY ARTERY

LEFT PULMONARY VEIN

LEFT PULMONARY VEIN

FAT

LAYER

PULMONARY VALVE

AORTIC VALVE

LEFT ATRIUM

MITRAL VALVE

CHORDAE TENDINEAE

RIGHT ATRIUM

TRICUSPID VALVE

INTERVENTRICULAR SEPTUM

LEFT VENTRICLE

MYOCARDIUM

LEFT CORONARY VEIN

LEFT CORONARY ARTERY

LEFT CORONARY NERVES

RIGHT VENTRICLE

DIUM

FAT

PERICARDIUM

AORTA

Pericardium
A thin, slippery bag, called the pericardium, wraps around the heart, enclosing a layer of fluid. This lubricates the heart's motions.

Mitral valve
This structure permits blood to go one way only. The long chordae tendineae help stop it flipping inside out.

▽ Heartbeat cycle

There are four phases to a heartbeat. The relax-and-refilling (phases 1 and 2 below) is called diastole. The contract-and-squeeze pumping (phases 3 and 4) is known as systole. The whole cycle lasts, on average, only four-fifths of a second. The four phases merge into one sequence.

Aortic valve shut

Pulmonary valve shut

Phase 1: the atriums relax and blood flows into them from the main veins.

Mitral valve open

Tricuspid valve open

Phase 2: blood passes from the atriums, through the tricuspid and mitral valves, to the ventricles below.

Mitral valve shut

Tricuspid valve shut

Phase 3: the ventricles contract and force blood through the aortic and pulmonary valves into the main arteries.

Aortic valve shut

Pulmonary valve shut

Phase 4: the ventricles have each ejected 70 ml (0.12 pint) of blood. They relax as the cycle begins again.

The Upper Back

HOLD YOUR ARMS STRAIGHT out sideways. Feel the muscles tensing in your upper arms and shoulders, and in your back, between your scapulae (shoulder blades). Before long, you will feel how heavy your arms and hands are. The broad, strong, intertwined muscles in your upper back and shoulders need to be strong enough to lift and move your arms and hands, as well as any objects you are holding.

Next, stretch both of your arms straight up in the air. Then hold them straight out in front of you, palms together; swing them back behind your head and finally let them hang by your sides. This demonstrates the wide range of movements made possible by your shoulders, one of your body's most mobile, flexible joints.

▷ A look back

The rib cage, scapulae (shoulder blades), and shoulder joints provide the framework for the powerful muscles of the upper back. Most joints are stable because the bones fit together snugly, held by ligaments. But in the shoulder, stability lies mainly in the muscles immediately around the bones: the deltoid, supraspinatus, infraspinatus, teres major, and teres minor.

This X-ray shows the three bones that meet at the shoulder joint – the scapula, clavicle, and humerus.

SEMISPINALIS CAPITIS

SPLENIUS CAPITIS

TRAPEZIUS

FIRST

SECOND RIB

THIRD

SUPRASPINATUS

ACROMION

SCAPULA

SEMISPINALIS

SUPRA SPINATUS

INFRA SPINATUS

JOINT CAPSULE

GLENOID CAVITY

HEAD OF HUMERUS

ARTERY OF SCAPULA

INFRASPINATUS

BICEPS

DELTOID

TERES MINOR

HUMERUS

TERES MAJOR

TRICEPS (LONG HEAD)

TERES MINOR

EIGHTH

NINTH RIB

TERES MAJOR

Under wraps

Smooth cartilage ("gristle") covers the ends of the humerus and scapula inside the joint to reduce friction as the joint moves.

THE SHOULDER JOINT

This is a ball-and-socket joint. The ball is the rounded top of the humerus bone, and the socket is a hollow in the scapula, called the glenoid cavity. To allow a wide range of movement, the socket is not as deep as the one in your hip bone, which is why the shoulder is more easily dislocated under stress.

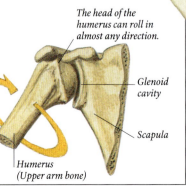

The head of the humerus can roll in almost any direction.

Glenoid cavity

Scapula

Humerus (Upper arm bone)

Shoulder holders

The supraspinatus, infraspinatus, and teres minor muscles support and protect the joint, keeping the head of the humerus bone in place.

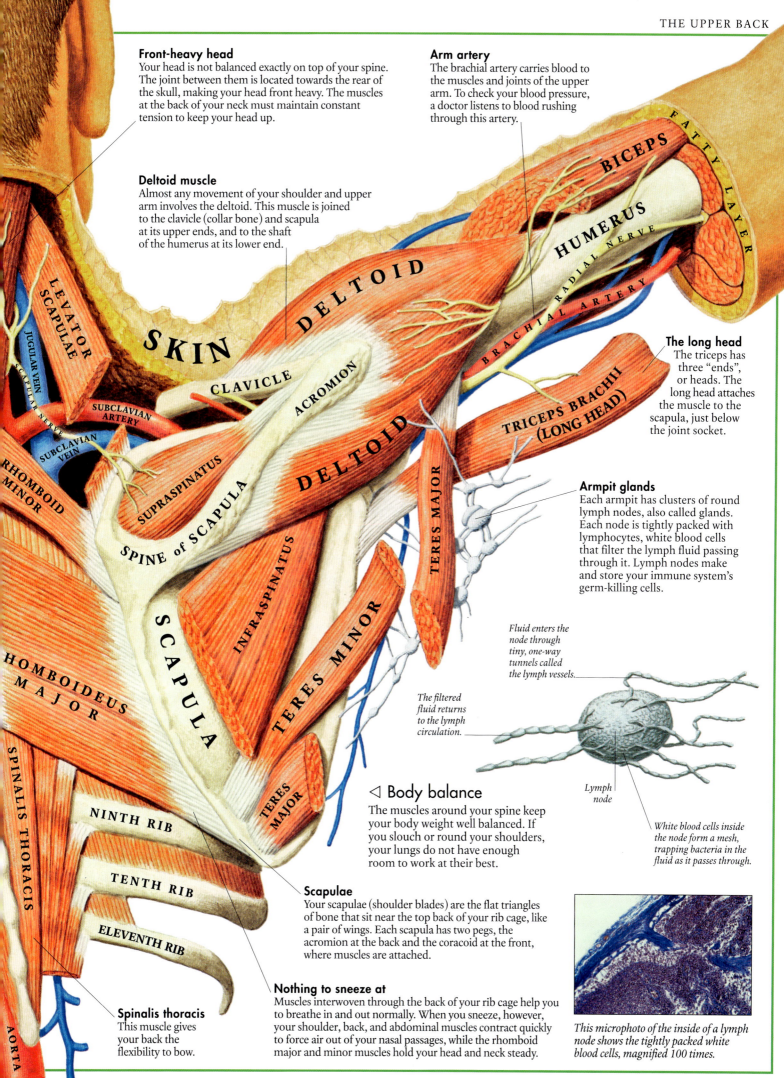

Front-heavy head
Your head is not balanced exactly on top of your spine. The joint between them is located towards the rear of the skull, making your head front heavy. The muscles at the back of your neck must maintain constant tension to keep your head up.

Deltoid muscle
Almost any movement of your shoulder and upper arm involves the deltoid. This muscle is joined to the clavicle (collar bone) and scapula at its upper ends, and to the shaft of the humerus at its lower end.

Arm artery
The brachial artery carries blood to the muscles and joints of the upper arm. To check your blood pressure, a doctor listens to blood rushing through this artery.

The long head
The triceps has three "ends", or heads. The long head attaches the muscle to the scapula, just below the joint socket.

Armpit glands
Each armpit has clusters of round lymph nodes, also called glands. Each node is tightly packed with lymphocytes, white blood cells that filter the lymph fluid passing through it. Lymph nodes make and store your immune system's germ-killing cells.

LEVATOR SCAPULAE

JUGULAR VEIN

SCAPULAR NERVE

SUBCLAVIAN ARTERY

SUBCLAVIAN VEIN

RHOMBOID MINOR

SKIN

DELTOID

CLAVICLE

ACROMION

SUPRASPINATUS

SPINE of SCAPULA

DELTOID

SCAPULA

INFRASPINATUS

RHOMBOIDEUS MAJOR

TERES MINOR

SPINALIS THORACIS

NINTH RIB

TENTH RIB

ELEVENTH RIB

TERES MAJOR

TERES MAJOR

BICEPS

FATTY LAYER

HUMERUS

RADIAL NERVE

BRACHIAL ARTERY

TRICEPS BRACHII (LONG HEAD)

Fluid enters the node through tiny, one-way tunnels called the lymph vessels.

The filtered fluid returns to the lymph circulation.

Lymph node

White blood cells inside the node form a mesh, trapping bacteria in the fluid as it passes through.

◁ **Body balance**
The muscles around your spine keep your body weight well balanced. If you slouch or round your shoulders, your lungs do not have enough room to work at their best.

Scapulae
Your scapulae (shoulder blades) are the flat triangles of bone that sit near the top back of your rib cage, like a pair of wings. Each scapula has two pegs, the acromion at the back and the coracoid at the front, where muscles are attached.

Nothing to sneeze at
Muscles interwoven through the back of your rib cage help you to breathe in and out normally. When you sneeze, however, your shoulder, back, and abdominal muscles contract quickly to force air out of your nasal passages, while the rhomboid major and minor muscles hold your head and neck steady.

Spinalis thoracis
This muscle gives your back the flexibility to bow.

AORTA

This microphoto of the inside of a lymph node shows the tightly packed white blood cells, magnified 100 times.

The Spinal Column

RUNNING DOWN THE MIDDLE OF YOUR BACK is the spine, a vital supporting rod for your head and body. The spine, also known as the spinal column or backbone, is a flexible chain of closely linked bones. These bones, called vertebrae, are linked by joints that allow slight movement with the bones above and below them. Over the entire length of the spine, however, these many small movements add up. Your spine lets you twist your upper body around, touch your toes, and turn a somersault.

The spinal column houses the spinal cord. This thick bundle of nerves transmits information back and forth between your brain and the rest of your body. It merges with the brain in the base of the skull, and extends partway down the inside of the spine through a tunnel of holes within the vertebrae. Muscles, blood vessels, and nerves sit at the front and sides of the spine.

△ Side view

Seen from the side, the spine has a gentle S-shaped curve, bending towards the rear in the neck and upper chest, and towards the front in the lower back. You have 24 individual vertebrae, of three main kinds: seven neck (cervical) vertebrae, twelve chest (thoracic) vertebrae, and five lower back (lumbar) vertebrae. Below these are the triangular sacrum and the tail-like coccyx.

Curved spine

Your spine has three main curves – cervical, lumbar, and thoracic – that help to balance the weight of your upper body evenly over your legs and feet. The cervical curve developed when you learned to hold up your own head as a baby, and the lumbar curve developed when you learned to walk and stand upright.

Cervical vertebrae

Side "wings" called transverse processes extend from each vertebra. The vertebral artery runs through holes in these processes. The seventh vertebra has a long backward-pointing hook, the spinous process, which anchors a neck ligament. You can feel it through your skin as a knob at the base of your neck.

Thoracic vertebrae

Each thoracic vertebra in the upper back supports a pair of ribs. Small hollows called facets hold the ribs in place. These bones work with the cervical vertebrae above when you bend over to tie your shoe, or bend back to look up.

A supple spine allows this acrobat to bend almost double.

▽ Rear view

The parts of your spine that stick out like hooks are called the spinous processes. They act as anchor points for muscles that keep your whole spine tensed and upright. The spinal cord does not continue to the base of your spinal column. It divides into numerous individual nerves at about the level of the first and second lumbar vertebrae.

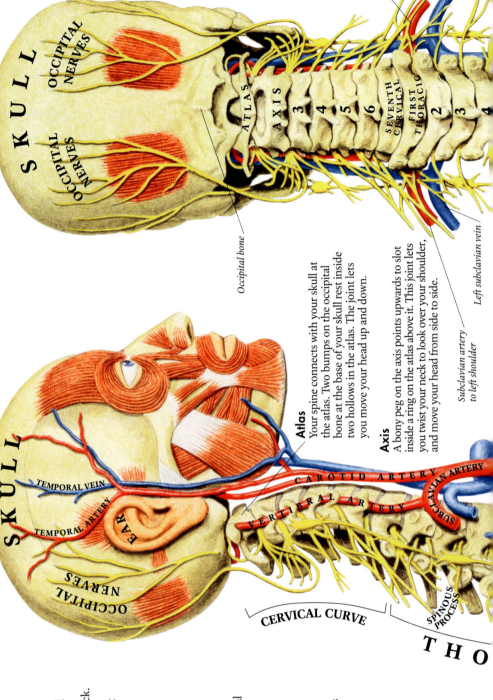

OCCIPITAL NERVES

OCCIPITAL NERVES

SKULL

ATLAS

AXIS

3
4
5
6

SEVENTH CERVICAL

FIRST THORACIC

2
3
4

Occipital bone

Right subclavian artery

Subclavian vein from right shoulder

Left subclavian vein

Atlas
Your spine connects with your skull at the atlas. Two bumps on the occipital bone at the base of your skull rest inside two hollows in the atlas. The joint lets you move your head up and down.

Axis
A bony peg on the axis points upwards to slot inside a ring on the atlas above it. This joint lets you twist your neck to look over your shoulder, and move your head from side to side.

SKULL

TEMPORAL VEIN

TEMPORAL ARTERY

EAR

OCCIPITAL NERVES

CAROTID ARTERY

VERTEBRAL ARTERY

SUBCLAVIAN ARTERY

Subclavian artery to left shoulder

CERVICAL CURVE

SPINOUS PROCESS

THO

THORACIC

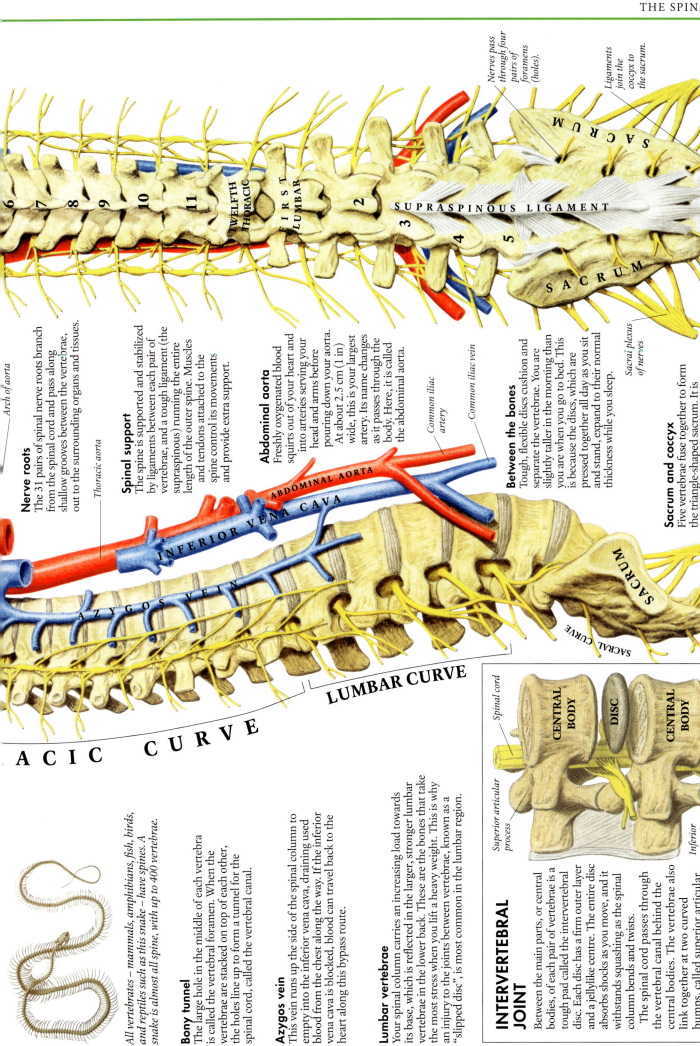

Nerves pass through four pairs of foramens (holes).

Ligaments join the coccyx to the sacrum.

SACRUM

6 7 8 9 10 11 TWELFTH THORACIC FIRST LUMBAR 2 3 4 5

SUPRASPINOUS LIGAMENT

SACRUM

Sacral plexus of nerves

Arch of aorta

Thoracic aorta

ABDOMINAL AORTA

INFERIOR VENA CAVA

AZYGOS VEIN

Common iliac artery

Common iliac vein

Spinal cord

CENTRAL BODY

DISC

CENTRAL BODY

Nerve roots

Superior articular process

Inferior articular facet

LUMBAR CURVE

SACRUM

COCCYX

SACRAL CURVE

ACIC CURVE

Nerve roots
The 31 pairs of spinal nerve roots branch from the spinal cord and pass along shallow grooves between the vertebrae, out to the surrounding organs and tissues.

Spinal support
The spine is supported and stabilized by ligaments between each pair of vertebrae, and a tough ligament (the supraspinous) running the entire length of the outer spine. Muscles and tendons attached to the spine control its movements and provide extra support.

Abdominal aorta
Freshly oxygenated blood squirts out of your heart and into arteries serving your head and arms before pouring down your aorta. At about 2.5 cm (1 in) wide, this is your largest artery. Its name changes as it passes through the body. Here, it is called the abdominal aorta.

Between the bones
Tough, flexible discs cushion and separate the vertebrae. You are slightly taller in the morning than you are when you go to bed. This is because the discs, which are pressed together all day as you sit and stand, expand to their normal thickness while you sleep.

Sacrum and coccyx
Five vertebrae fuse together to form the triangle-shaped sacrum. It is joined to the bottom lumbar vertebra above, the hip bones on either side, and the coccyx below. The coccyx, or "tail bone", contains about four fused vertebrae.

All vertebrates – mammals, amphibians, fish, birds, and reptiles such as this snake – have spines. A snake is almost all spine, with up to 400 vertebrae.

Bony tunnel
The large hole in the middle of each vertebra is called the vertebral foramen. When the vertebrae are stacked on top of each other, the holes line up to form a tunnel for the spinal cord, called the vertebral canal.

Azygos vein
This vein runs up the side of the spinal column to empty into the inferior vena cava, draining used blood from the chest along the way. If the inferior vena cava is blocked, blood can travel back to the heart along this bypass route.

Lumbar vertebrae
Your spinal column carries an increasing load towards its base, which is reflected in the larger, stronger lumbar vertebrae in the lower back. These are the bones that take the most stress when you lift a heavy weight. This is why an injury to the joints between vertebrae, known as a "slipped disc", is most common in the lumbar region.

INTERVERTEBRAL JOINT
Between the main parts, or central bodies, of each pair of vertebrae is a tough pad called the intervertebral disc. Each disc has a firm outer layer and a jellylike centre. The entire disc absorbs shocks as you move, and it withstands squashing as the spinal column bends and twists.
The spinal cord passes through the vertebral canal behind the central bodies. The vertebrae also link together at two curved bumps, called superior articular processes, which fit into inferior articular facets of the vertebra just above, to form two joints.

The Arm and Hand

YOU ARE WALKING THROUGH A GARDEN at night, guided by the light from your torch. Suddenly, it fails, and you find yourself in pitch darkness. Instinctively, you stretch out your arms in front of you and fan your fingertips, groping for the slightest contact. Your arms and hands are important for their touching and sensing abilities, as well as for their ability to grasp and manipulate.

As you feel your way toward the light of the distant house, one of your fingers touches a needle-like point. You react almost at once, by jerking your hand away. Then your torch comes on again, and you can see the point – a prickly thistle. The skin in your fingertip registered the pressure of a thorn. Nerve signals flashed the information to your brain, your brain sent out nerve signals to the many muscles in your arm, and these muscles contracted, pulling your hand away from the thistle – all in a split second.

◁ Juggling hands

As the expert juggler throws and catches, his eyes follow the movements of the balls. Muscles in his arms and hands make them reach out to the exact position where each ball will fall. The skin on his hands confirms contact and safe grip of the ball, even as the muscles prepare for the next throw. It is an amazing series of fast, precise movements.

▽ A sensitive touch

Human arms and hands are incredibly flexible. Their complex network of nerves, blood vessels, and muscles makes them among the most efficient manipulating devices in nature. Nerve endings and clusters of sense receptors in the skin, particularly in extra-sensitive areas such as the fingertips, send messages to the brain. Strong tendons joining the muscles to the bones mean that the body can react to these messages very quickly.

Wristbands

About 10 blood vessels and nerves, and more than 20 tendons, pass through the wrist area. These are bound by two fibrous bands just under the skin, which together look like a wide watch strap. The bands are the flexor retinaculum on the palm side (not visible here), and the extensor retinaculum on the back of the wrist, here separated to expose the tendons running under it.

The first digit, or finger, is known as the index finger. Index comes from a Latin word meaning "pointer".

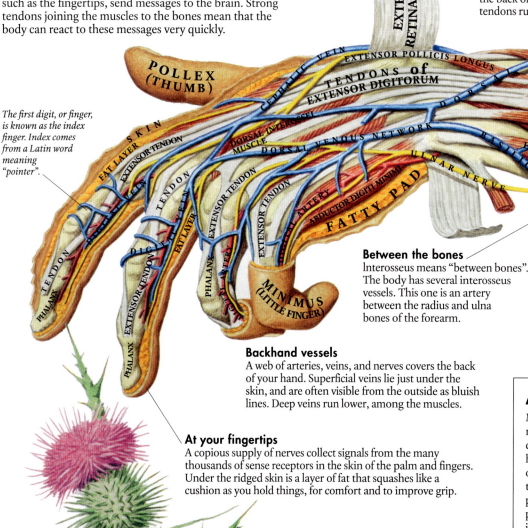

Between the bones

Interosseus means "between bones". The body has several interosseus vessels. This one is an artery between the radius and ulna bones of the forearm.

Twist of the wrist

This is the muscle you use to bend your hand at the wrist or twist your wrist. It gets its name from the carpus (wrist bone) and the ulna (forearm bone).

Backhand vessels

A web of arteries, veins, and nerves covers the back of your hand. Superficial veins lie just under the skin, and are often visible from the outside as bluish lines. Deep veins run lower, among the muscles.

At your fingertips

A copious supply of nerves collect signals from the many thousands of sense receptors in the skin of the palm and fingers. Under the ridged skin is a layer of fat that squashes like a cushion as you hold things, for comfort and to improve grip.

A SENSE OF POSITION

Many of the body's muscles and joints have microscopic stretch sensors in them. They detect whether a muscle is pulled tight or hanging loose, and whether a joint is bent or straight. Nerves send this information to the brain, which can then work out the positions and postures of various body parts. This awareness of body position is called the proprioceptive sense.

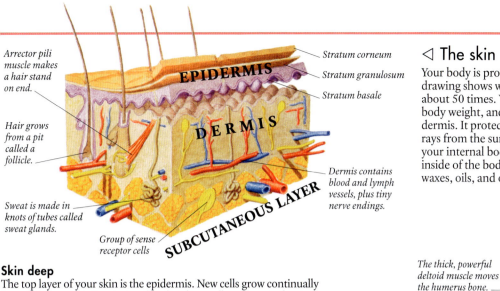

Arrector pili muscle makes a hair stand on end.

EPIDERMIS

Stratum corneum

Stratum granulosum

Stratum basale

DERMIS

Hair grows from a pit called a follicle.

Dermis contains blood and lymph vessels, plus tiny nerve endings.

Sweat is made in knots of tubes called sweat glands.

Group of sense receptor cells

SUBCUTANEOUS LAYER

◁ The skin

Your body is protected and covered by its outer layer, the skin. This drawing shows what a section of your skin looks like, magnified about 50 times. Your skin makes up more than one-tenth of your body weight, and is made of two parts, the epidermis and the dermis. It protects your internal organs, helping to stop harmful rays from the sun and invading germs from reaching them. It stops your internal body fluids from leaking away and helps to keep the inside of the body at a constant temperature. Your skin also contains waxes, oils, and other substances that make it waterproof.

Skin deep

The top layer of your skin is the epidermis. New cells grow continually from its lower part, the stratum basale. These cells pass up to the stratum granulosum. There, some cells make the protein keratin, which toughens the cells. Other cells in the stratum basale produce melanin, the substance that gives skin its colour. The stratum granulosum cells fill with keratin and die as they continue upward to reach the top surface, the stratum corneum. After a month or so, the cells are worn away by friction and replaced. The dermis, about four times thicker than the epidermis, consists mainly of the protein collagen, which builds scar tissue to mend cuts and abrasions. Underneath it is the fatty subcutaneous layer.

Biceps and triceps

The biceps and triceps muscles control the up and down movements of your forearm. The biceps has two heads, or points of attachment to the bone. The triceps has three heads: long, lateral, and medial. To distinguish between the biceps muscle in your leg (the biceps femoris muscle), the one in your arm is called the biceps brachii.

The thick, powerful deltoid muscle moves the humerus bone.

DELTOID MUSCLE

SCAPULA

TERES MINOR

TERES MAJOR

HUMERUS

LONG HEAD

SHORT HEAD

CEPHALIC VEIN

BICEPS BRACHII

BRACHIALIS

TRICEPS LATERAL HEAD

BRACHIAL ARTERY

RADIAL NERVE

LONG HEAD

TRICEPS

CEPHALIC VEIN

DIGITORUM

BRACHIORADIALIS

EXTENSOR CARPI RADIALIS LONGUS

EXTENSOR CARPI ULNARIS

TRICEPS LATERAL HEAD

FATTY LAYER

LATISSIMUS DORSI

RIB

RIB

ANCONEUS

TRICEPS TENDON

SKIN

SERRATUS ANTERIOR

ULNA

MEDIAN VEIN

ELBOW

A bypass for blood

The brachial artery is the main tube bringing fresh blood into the arm. Just above the elbow, it splits to form the radial and ulnar arteries – the chief forearm vessels. The brachial artery also has several smaller branches that divide and join again, to create a network of bypasses or "short cuts" for the blood. This type of system is called collateral circulation. Blood can get through via a number of collateral, or side-by-side, routes.

HAND TOUCH MAP

The skin on the palms of your hands – especially on your fingertips – is richly supplied with sense receptors that provide your sense of touch. On the back of your hand, as in many parts of the body, these receptors are wrapped round the base of a hair, so that they can sense any movement of the hair shaft. In hairless areas, like the palms, lips, and tongue, the cells are thickly clustered in disks within the skin.

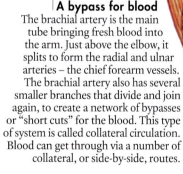

The darker the shading, the more sensitive the area.

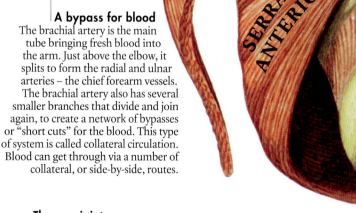

Less sensitive   More sensitive

The arm joints

The series of joints from the shoulder to the fingertips make your arms extremely flexible. The ball-and-socket joint inside your shoulder lets you move your arm in almost any direction. The hinge joint at your elbow enables you to fold your arm in half and straighten it out again. The joint at the junction of the wrist bones gives flexibility to your hand. You can bend and stretch your fingers because of the hinge joints inside them.

The Shoulder and Elbow

PRETEND YOU'VE WON! Raise your clenched fist in a salute of triumph. You are using many of the powerful muscles shown here, in your shoulder and upper arm. The bulge you see under the skin of your upper arm is the biceps brachii muscle. Like other muscles, the biceps bulges in the middle as it shortens. This type of muscle action is called isotonic contraction. The muscle keeps the same tone, or pulling power, but shortens as the elbow bends. Now clench your fist and tense your biceps like a bodybuilder, but keep your arm still. The biceps still bulges. This is isometric contraction: the muscle exerts more pulling power, but stays the same length. Most muscles use a combination of isometric and isotonic contractions to move or hold steady the parts of your body.

Biceps and triceps muscles work together to stabilize the arm.

This gymnast working on a pommel horse can support his entire body weight using the muscles of the upper arm and shoulder.

▷ The shoulder and upper arm

Whether you are lifting a heavy weight or picking up a feather, the muscles in your upper arm and shoulder provide both power and accuracy. The muscles in the upper back and shoulder move the upper arm; muscles in the upper arm move the forearm; and muscles in the forearm move the wrist and hand. The whole arm is a three-section lever. With it, you can reach out and grasp an apple, then fold your arm back on itself to raise the apple to your mouth.

THE "DOUBLE-JOINTED" SHOULDER

Your shoulder is a very flexible ball-and-socket joint. A significant factor in its mobility is the movement of the scapula (shoulder blade). Its movements alter the angle of the socket, which cradles the ball at the upper end of the humerus (upper arm bone). When you raise your arm through a full half-circle (180°) – from hanging at your side to straight above your head – about half of this movement is due to the scapula shifting its position.

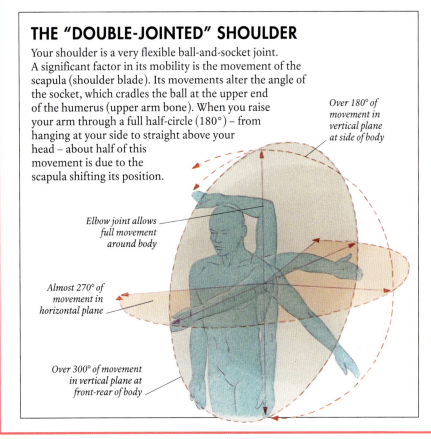

Over 180° of movement in vertical plane at side of body

Elbow joint allows full movement around body

Almost 270° of movement in horizontal plane

Over 300° of movement in vertical plane at front-rear of body

Pectoralis pairs
Under the thick triangular slab of the pectoralis major, the main front chest muscle, is the smaller pectoralis minor. It links the upper ribs with the scapula and the humerus. When it contracts, it pulls the scapula around to the side of the rib cage. This allows your arm to swing across your chest, so that your fingers can touch the opposite shoulder.

Rib muscles
Thin sheets of muscle called the intercostals weave between the bones of your rib cage. Nerves and blood vessels also fill the intercostal space between the ribs.

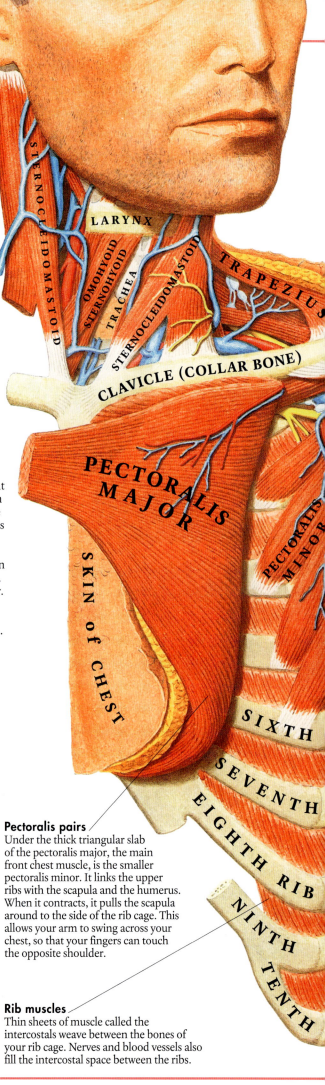

STERNOCLEIDOMASTOID

LARYNX

OMOHYOID

STERNOHYOID

TRACHEA

STERNOCLEIDOMASTOID

TRAPEZIUS

CLAVICLE (COLLAR BONE)

PECTORALIS MAJOR

PECTORALIS MINOR

SKIN OF CHEST

SIXTH

SEVENTH

EIGHTH RIB

NINTH

TENTH

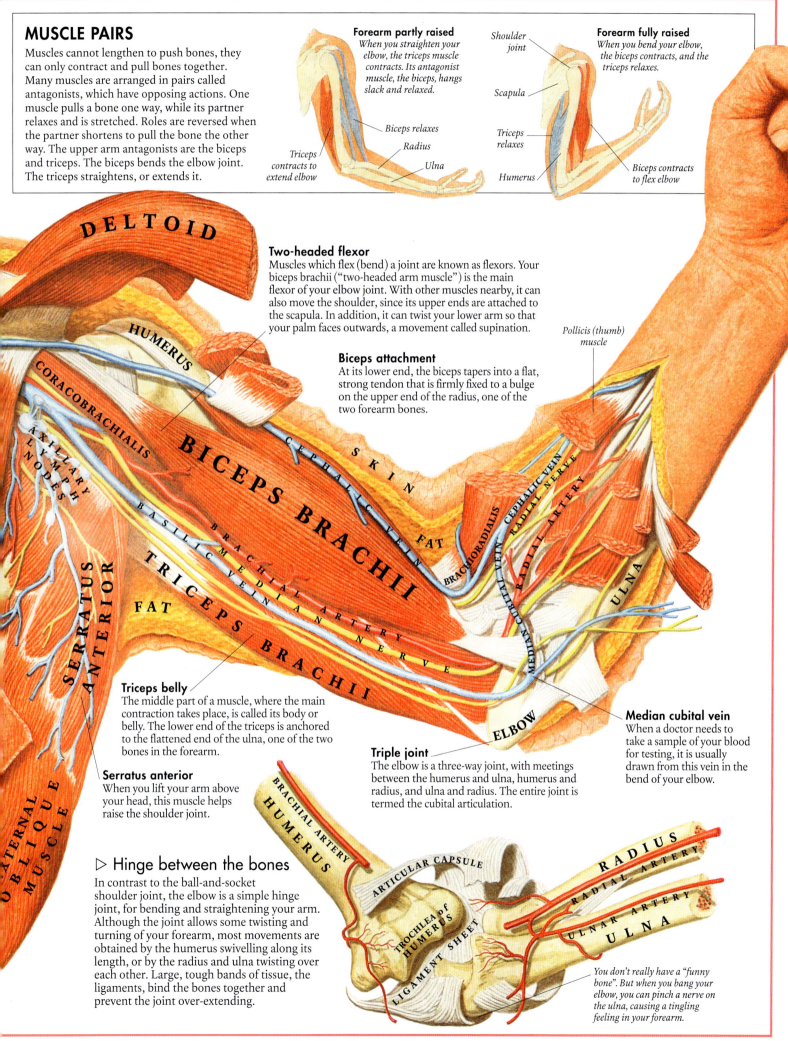

MUSCLE PAIRS

Muscles cannot lengthen to push bones, they can only contract and pull bones together. Many muscles are arranged in pairs called antagonists, which have opposing actions. One muscle pulls a bone one way, while its partner relaxes and is stretched. Roles are reversed when the partner shortens to pull the bone the other way. The upper arm antagonists are the biceps and triceps. The biceps bends the elbow joint. The triceps straightens, or extends it.

Forearm partly raised
When you straighten your elbow, the triceps muscle contracts. Its antagonist muscle, the biceps, hangs slack and relaxed.

Biceps relaxes
Radius
Ulna
Triceps contracts to extend elbow

Shoulder joint
Scapula
Triceps relaxes
Humerus

Forearm fully raised
When you bend your elbow, the biceps contracts, and the triceps relaxes.

Biceps contracts to flex elbow

DELTOID
HUMERUS
CORACOBRACHIALIS
AXILLARY LYMPH NODES
BICEPS BRACHII
CEPHALIC VEIN
SKIN
FAT
BASILIC VEIN
BRACHIAL ARTERY
MEDIAN NERVE
MEDIAL
SERRATUS ANTERIOR
TRICEPS BRACHII
FAT
EXTERNAL OBLIQUE MUSCLE

BRACHORADIALIS
CEPHALIC VEIN
RADIAL NERVE
MEDIAN CUBITAL VEIN
RADIAL ARTERY
ULNA
ELBOW

Pollicis (thumb) muscle

Two-headed flexor

Muscles which flex (bend) a joint are known as flexors. Your biceps brachii ("two-headed arm muscle") is the main flexor of your elbow joint. With other muscles nearby, it can also move the shoulder, since its upper ends are attached to the scapula. In addition, it can twist your lower arm so that your palm faces outwards, a movement called supination.

Biceps attachment

At its lower end, the biceps tapers into a flat, strong tendon that is firmly fixed to a bulge on the upper end of the radius, one of the two forearm bones.

Triceps belly

The middle part of a muscle, where the main contraction takes place, is called its body or belly. The lower end of the triceps is anchored to the flattened end of the ulna, one of the two bones in the forearm.

Serratus anterior

When you lift your arm above your head, this muscle helps raise the shoulder joint.

Triple joint

The elbow is a three-way joint, with meetings between the humerus and ulna, humerus and radius, and ulna and radius. The entire joint is termed the cubital articulation.

Median cubital vein

When a doctor needs to take a sample of your blood for testing, it is usually drawn from this vein in the bend of your elbow.

▷ Hinge between the bones

In contrast to the ball-and-socket shoulder joint, the elbow is a simple hinge joint, for bending and straightening your arm. Although the joint allows some twisting and turning of your forearm, most movements are obtained by the humerus swivelling along its length, or by the radius and ulna twisting over each other. Large, tough bands of tissue, the ligaments, bind the bones together and prevent the joint over-extending.

BRACHIAL ARTERY
HUMERUS
ARTICULAR CAPSULE
TROCHLEA of HUMERUS
LIGAMENT SHEET
RADIUS
RADIAL ARTERY
ULNAR ARTERY
ULNA

You don't really have a "funny bone". But when you bang your elbow, you can pinch a nerve on the ulna, causing a tingling feeling in your forearm.

The Wrist, Hand, and Fingers

YOUR HANDS ARE SENSITIVE enough to thread cotton through the eye of a needle, yet strong enough to squeeze and crush. These abilities depend not only on your hands, but also on your brain, which controls them through your nervous system, and on your senses, which provide your brain with the necessary information.

Each hand is built around a skeletal framework of 27 bones, connected by a multitude of complex joints that allow amazing flexibility. Wrapped around the bones are muscles and their tendons, blood vessels, nerves – and very little else. Enclosing them all is a layer of skin that contains some of the most sensitive patches in your body, especially on your fingertips. This skin also bears a ridged pattern of swirls and whorls that makes you unique among the 7-billion-plus humans on Earth – your fingerprints.

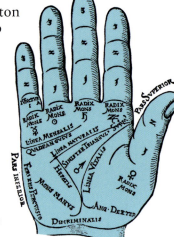

◁ Reading the lines

Like fingerprints, the pattern of creases and lines in the skin of the palm is different for each person. Some people believe that the palm pattern reveals the history, health, and fate of the owner. Interpreting creases and lines is called chiromancy, or palmistry. Some palmists claim they can get an impression of the person by chirognomy, studying the hand's overall shape, colour, texture, and flexibility.

A chart from the 16th century shows lines and other features interpreted by a palmist.

MECHANICAL HANDS

Engineers have tried for many years to perfect an artificial hand that can mimic the complicated movements of a human hand. Jointed robotic arms and hands carry out repetitive tasks in factory assembly lines and laboratories, and do dangerous work such as bomb disposal and space exploration. Artificial replacement (prosthetic) hands for people who have lost theirs are improving all the time and are now able to carry out many of the functions of the original limb. There are versions in development that can be connected directly to the brain (brain-machine interfacing) to allow control of movement and perception of touch and temperature.

Individually controlled "fingers" give this robot its grip.

▷ In the palm of your hand

In the view on the right, the skin of the palm is peeled back to reveal the structures that make your hand so flexible. From the thumb (or pollex, meaning "strong") to the little finger (or minimus, meaning "least"), the finger bones are crossed by tendons wrapped in long sheaths. These link them to strong muscles in the forearm which control most finger movements. The muscles attached to the metacarpal bones in the palm provide more pulling power, and also give shape to your hand. Blood reaches right up to your fingertips along branches of the radial and ulnar arteries. The entire area is well supplied with nerve endings, linked to the radial, ulnar, and median nerves.

▽ Bones of the hand

Your hand has three main anatomical regions. These are the carpus (wrist), the metacarpus (palm), and the digits (fingers). The carpus has eight bones, in two rows of four. Four of these bones link with the radius and ulna bones. Five metacarpal bones stretch from your wrist to your knuckles, and 14 phalanx bones shape your fingers. About 40 ligaments strap the bones together, most of these in the wrist.

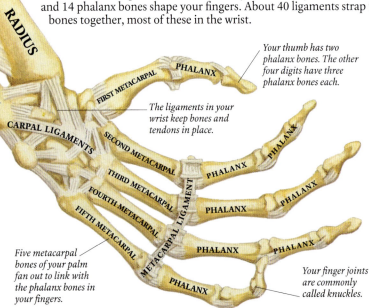

RADIUS

ULNA

CARPAL LIGAMENTS

FIRST METACARPAL

PHALANX

SECOND METACARPAL

THIRD METACARPAL

FOURTH METACARPAL

FIFTH METACARPAL

METACARPAL LIGAMENT

PHALANX

PHALANX

PHALANX

PHALANX

PHALANX

PHALANX

PHALANX

PHALANX

Your thumb has two phalanx bones. The other four digits have three phalanx bones each.

The ligaments in your wrist keep bones and tendons in place.

Five metacarpal bones of your palm fan out to link with the phalanx bones in your fingers.

Your finger joints are commonly called knuckles.

GRIPPING STUFF

Your hand is well-designed for gripping objects. These three types of grip demonstrate the flexibility of the human hand. The precision grip allows you to hold an object delicately between forefinger and thumb. No other animals, even our close cousins, chimps and gorillas, have this type of grip. The thumb can also be tilted to touch, or oppose, the other fingertips. You can mould your fingers and palm around an object with a spherical grip, giving you a secure hold. The power grip of two hands can be strong enough to hold the body's entire weight – imagine hanging on to the edge of a cliff!

A precision grip allows highly co-ordinated work, but does not provide a secure hold.

A spherical grip is used when holding a round object. The width of the palm gives stability.

In a power grip, the fingers wrap round the object held, with added pressure from the thumb.

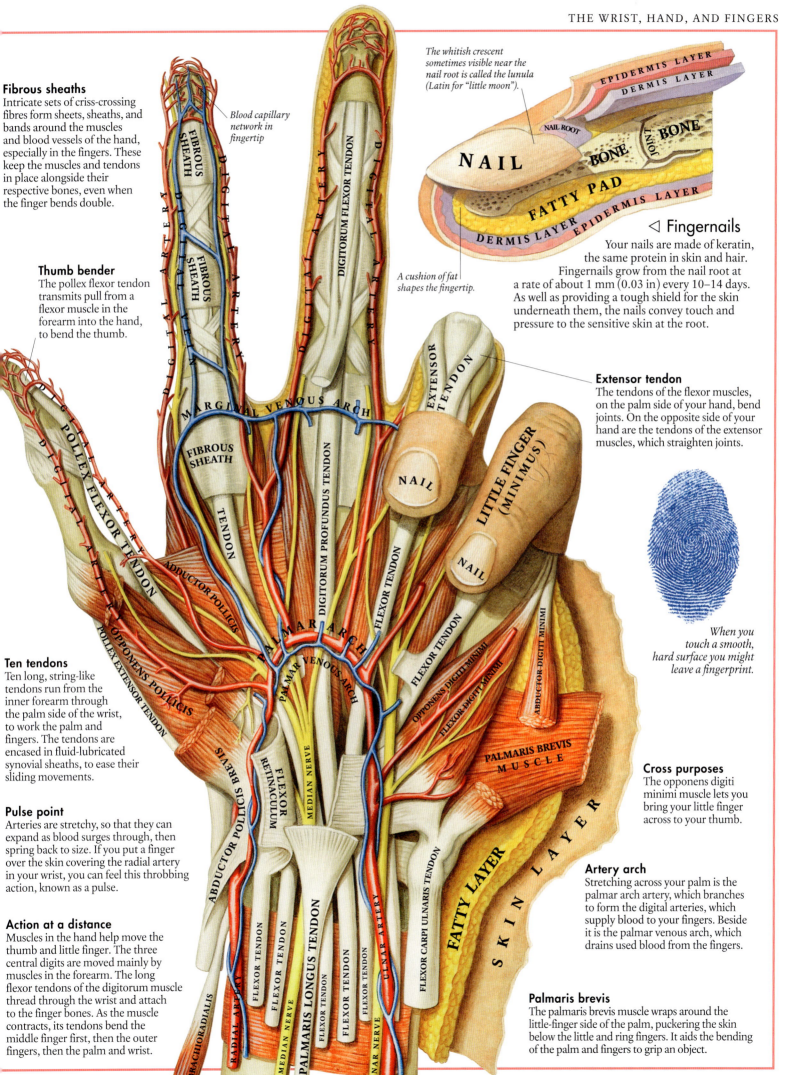

Fibrous sheaths
Intricate sets of criss-crossing fibres form sheets, sheaths, and bands around the muscles and blood vessels of the hand, especially in the fingers. These keep the muscles and tendons in place alongside their respective bones, even when the finger bends double.

Thumb bender
The pollex flexor tendon transmits pull from a flexor muscle in the forearm into the hand, to bend the thumb.

Ten tendons
Ten long, string-like tendons run from the inner forearm through the palm side of the wrist, to work the palm and fingers. The tendons are encased in fluid-lubricated synovial sheaths, to ease their sliding movements.

Pulse point
Arteries are stretchy, so that they can expand as blood surges through, then spring back to size. If you put a finger over the skin covering the radial artery in your wrist, you can feel this throbbing action, known as a pulse.

Action at a distance
Muscles in the hand help move the thumb and little finger. The three central digits are moved mainly by muscles in the forearm. The long flexor tendons of the digitorum muscle thread through the wrist and attach to the finger bones. As the muscle contracts, its tendons bend the middle finger first, then the outer fingers, then the palm and wrist.

Blood capillary network in fingertip

The whitish crescent sometimes visible near the nail root is called the lunula (Latin for "little moon").

A cushion of fat shapes the fingertip.

◁ Fingernails
Your nails are made of keratin, the same protein in skin and hair. Fingernails grow from the nail root at a rate of about 1 mm (0.03 in) every 10–14 days. As well as providing a tough shield for the skin underneath them, the nails convey touch and pressure to the sensitive skin at the root.

Extensor tendon
The tendons of the flexor muscles, on the palm side of your hand, bend joints. On the opposite side of your hand are the tendons of the extensor muscles, which straighten joints.

When you touch a smooth, hard surface you might leave a fingerprint.

Cross purposes
The opponens digiti minimi muscle lets you bring your little finger across to your thumb.

Artery arch
Stretching across your palm is the palmar arch artery, which branches to form the digital arteries, which supply blood to your fingers. Beside it is the palmar venous arch, which drains used blood from the fingers.

Palmaris brevis
The palmaris brevis muscle wraps around the little-finger side of the palm, puckering the skin below the little and ring fingers. It aids the bending of the palm and fingers to grip an object.

The Lower Torso

YOUR LOWER TORSO contains the machinery for metabolism, the name for the thousands of chemical processes in the body. The digestive system, consisting mainly of stomach and intestines, takes up most of the space in your abdomen. These organs lie coiled together within your torso, digesting and absorbing energy- and nutrient-rich substances for growth and repair. Although food is inside your body once you swallow it, it is not truly a part of your body until it is absorbed through the lining of your stomach and intestines, into your body tissues.

Your liver, wedged into the top of the lower torso, is like a chemical factory, carrying out hundreds of tasks within its two lobes. The bottom of your lower torso contains two exits. Your digestive system expels undigested food and other left-overs from the intestines through the anus, and your urinary system gets rid of wastes filtered from your blood through the urethra.

▷ A maze of tubes

Moving the intestines outward and slightly to the side exposes the intricate network of arteries and veins supplying blood to the abdominal organs. Normally, these are hidden by the coils of the small and large intestines. Behind this maze of piping, against the rear of the abdominal wall, sit the kidneys.

▽ Main organs of the abdomen

If you peeled away the skin and fat at the front of the abdomen, you would see how the organs are coiled, curled, and packed inside, forming a compact conveyor belt for metabolism. The abdomen is bounded above by the dome-shaped diaphragm; at the rear by the spine, hip bones, and back muscles; to the sides and front by large muscle-and-fibre sheets in the abdominal wall, and at the base by the pelvis.

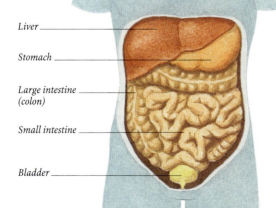

Chest (thorax)

Liver

Stomach

Large intestine (colon)

Small intestine

Bladder

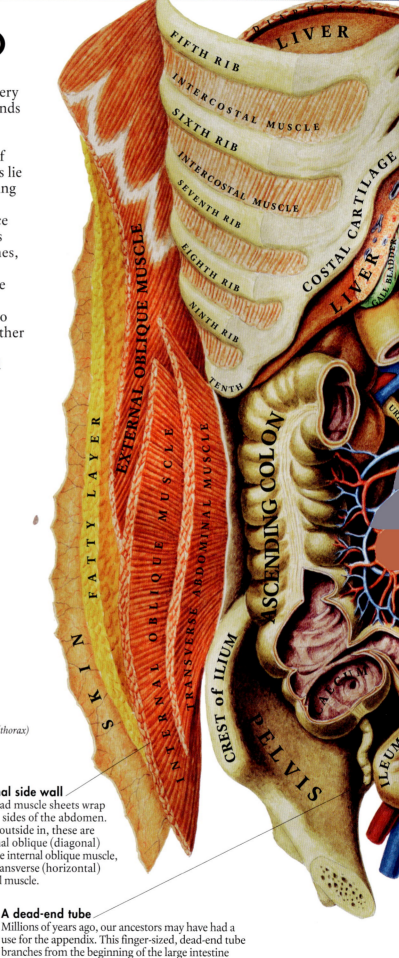

Abdominal side wall

Three broad muscle sheets wrap round the sides of the abdomen. From the outside in, these are the external oblique (diagonal) muscle, the internal oblique muscle, and the transverse (horizontal) abdominal muscle.

A dead-end tube

Millions of years ago, our ancestors may have had a use for the appendix. This finger-sized, dead-end tube branches from the beginning of the large intestine (colon). Some animals that eat plant food, such as rabbits, have a proportionally larger appendix that helps them to digest food. In humans, it's not certain whether the appendix still has a function, but it can be removed without ill effects if it becomes inflamed (appendicitis).

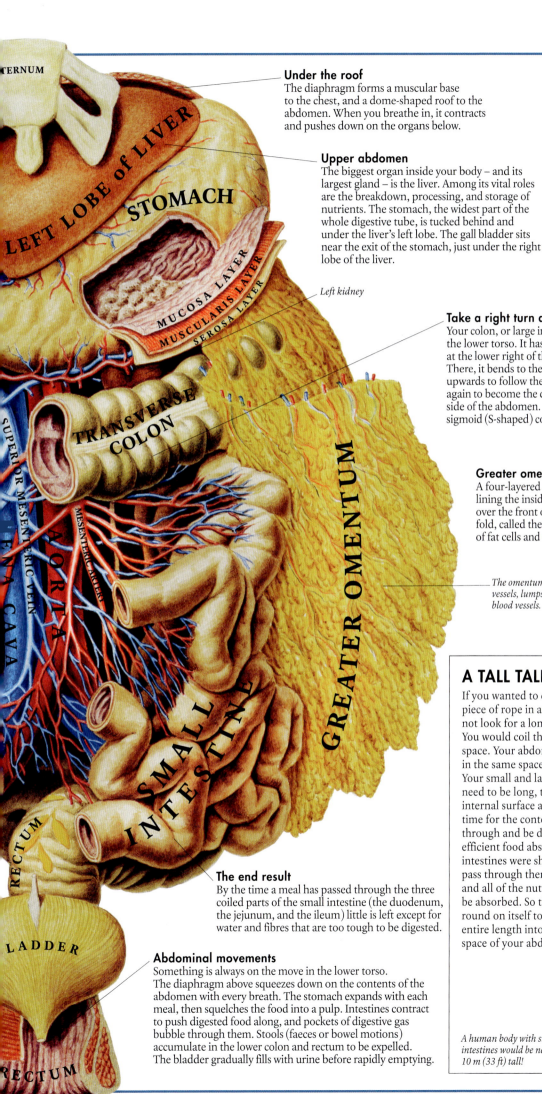

Under the roof

The diaphragm forms a muscular base to the chest, and a dome-shaped roof to the abdomen. When you breathe in, it contracts and pushes down on the organs below.

Upper abdomen

The biggest organ inside your body – and its largest gland – is the liver. Among its vital roles are the breakdown, processing, and storage of nutrients. The stomach, the widest part of the whole digestive tube, is tucked behind and under the liver's left lobe. The gall bladder sits near the exit of the stomach, just under the right lobe of the liver.

Left kidney

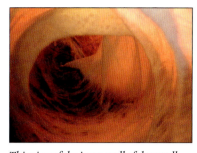

This view of the inner wall of the small intestine was taken with an endoscope, a tube-like instrument for looking into body cavities. Endoscopes can see into the gullet, stomach, small intestine, large intestine, and bladder.

Take a right turn at the liver

Your colon, or large intestine, makes an arch as it curls around the lower torso. It has four sections. The ascending colon starts at the lower right of the abdomen and stretches up to the liver. There, it bends to the left to form the transverse colon. It passes upwards to follow the curve of the stomach, then bends down again to become the descending colon, which runs down the left side of the abdomen. Near the top of the pelvis, it forms the sigmoid (S-shaped) colon, curving to connect with the rectum.

Greater omentum

A four-layered fold of peritoneum (the membrane lining the inside of the abdomen) hangs down over the front of the intestines like an apron. This fold, called the greater omentum, contains pads of fat cells and infection-fighting lymph nodes.

The omentum contains lymph vessels, lumps of fat, and some blood vessels.

The end result

By the time a meal has passed through the three coiled parts of the small intestine (the duodenum, the jejunum, and the ileum) little is left except for water and fibres that are too tough to be digested.

Abdominal movements

Something is always on the move in the lower torso. The diaphragm above squeezes down on the contents of the abdomen with every breath. The stomach expands with each meal, then squelches the food into a pulp. Intestines contract to push digested food along, and pockets of digestive gas bubble through them. Stools (faeces or bowel motions) accumulate in the lower colon and rectum to be expelled. The bladder gradually fills with urine before rapidly emptying.

A TALL TALE

If you wanted to carry a long piece of rope in a bag, you would not look for a long, thin bag. You would coil the rope, to save space. Your abdomen is designed in the same space-saving way. Your small and large intestines need to be long, to provide a large internal surface area and enough time for the contents to pass through and be digested, for efficient food absorption. If your intestines were short, food would pass through them too quickly, and all of the nutrients might not be absorbed. So the intestine coils round on itself to squeeze its entire length into the compact space of your abdomen.

Oesophagus

Stomach

The small intestine is about 5–6 m (16–20 ft) long.

The large intestine is about 1.5 m (5 ft) long.

Rectum

A human body with straight intestines would be nearly 10 m (33 ft) tall!

The Stomach

HAVE YOU EVER EATEN so much that you felt you were about to burst? If so, your stomach had probably stretched to hold around two litres (nearly half a gallon) of contents. Perhaps you complained of "stomach ache" and pointed to your navel area. In fact, the stomach is much higher in the body than most people imagine. It is tucked under your lower left ribs, with its base about level with the lowest rib.

Your stomach is the second stop for food and drink, after your mouth. A J-shaped bag made of several muscle layers, its main job is to break food into smaller pieces for digestion by squeezing and squelching it into a sloppy mush. It also breaks down food chemically, by mixing it with acid and digestive chemicals called gastric enzymes, all made in the stomach lining. Not many germs can withstand this chemical assault, so the stomach also helps to sterilize the food you eat.

FILLING AND EMPTYING

An average meal takes about six hours to pass through the stomach on its way to the small intestine. Starchy, carbohydrate-rich foods are digested in two or three hours, high-protein foods take slightly longer, and fatty foods may still be trickling from the stomach seven or eight hours after you eat them.

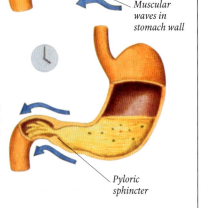

A meal reaches your stomach as a series of soft, saliva moistened balls called boluses. With each swallow, food accumulates in your stomach, making it expand like a balloon.

After an hour or two, the food has been mashed and mixed with acid and enzymes to form a creamy liquid called chyme. It is ready to be squeezed into the small intestine.

Muscular waves in stomach wall

A few hours later, some food has oozed through the stomach's exit, the pyloric sphincter, but most is still churning in the stomach. The stomach does not stay large and rigid – it gradually shrinks, like a deflating balloon.

Pyloric sphincter

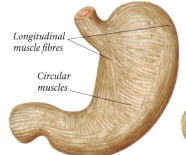

Longitudinal muscle fibres

Circular muscles

△ Outer muscle layers

Under its smooth outer coat, the serosa, the stomach has several muscle layers. Two groups of muscle fibres, the longitudinals, run down its sides. Under these are circular muscles, around the entire stomach.

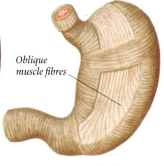

Oblique muscle fibres

△ Inner muscle layers

Beneath the circular muscles are diagonal bands of muscle fibres, the obliques. Overall, the three sets of criss-crossing muscles let the stomach contract in almost any direction, to squeeze its contents.

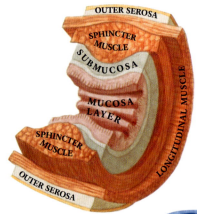

OUTER SEROSA
SPHINCTER MUSCLE
SUBMUCOSA
MUCOSA LAYER
SPHINCTER MUSCLE
OUTER SEROSA
LONGITUDINAL MUSCLE

◁ Pyloric sphincter

The gate between the stomach and the small intestine is the pyloric sphincter. It is formed from a ring-like thickening in the circular band of muscle. Usually closed to hold in the stomach contents, it relaxes for a few seconds at a time during digestion. Pressure from the squeezing stomach muscles forces a squirt of chyme through the sphincter into the intestine.

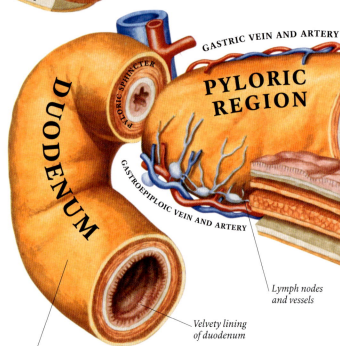

GASTRIC VEIN AND ARTERY
PYLORIC REGION
PYLORIC SPHINCTER
DUODENUM
GASTROEPIPLOIC VEIN AND ARTERY
Lymph nodes and vessels
Velvety lining of duodenum

A big push

The narrow, tube-like section of the stomach nearest to the first part of the small intestine, the duodenum, is known as the pyloric region. When food is ready to leave the stomach and enter the duodenum, the muscles there squeeze in forceful waves, called peristaltic contractions, to push out the food.

Arteries and veins

Like all organs, the stomach has its own blood supply. A team of arteries branching from the abdominal aorta brings it fresh, high-oxygen blood. Veins return most of the used, low-oxygen blood back to the heart. But some blood travels instead along branches of the portal vein to the liver. There, substances in the blood absorbed through the stomach wall, such as sugars and alcohols, are processed.

OESOPHAGUS (GULLET)

BOLUS

MUCOSA LAYER
SUBMUCOSA LAYER
MUSCLE LAYERS
SEROSA LAYER

The speed of a swallow
Like the stomach and other parts of the digestive tube, the oesophagus has layers of longitudinal and circular muscles in its wall. These muscles contract in waves to squeeze a food bolus down towards the stomach at a rate of about 4 cm (1.5 in) per second.

Not a true sphincter
The junction of the oesophagus and stomach is sometimes called the cardiac sphincter because it is near to the heart. Although it cannot open and close as tightly as the pyloric sphincter, the cardiac sphincter is aided by the contraction of other muscles to keep acid and swallowed food from welling back up the oesophagus (a feeling known as "heartburn").

Fundus of stomach
This confusing name for the top of the stomach actually means "bottom". In an operation, the stomach is often opened from beneath, so the fundus would appear to be at the bottom.

FUNDUS of STOMACH

GASTRIC ARTERY
GASTRIC VEIN

LYMPH NETWORK

INNER STOMACH

CHYME
SEMI-DIGESTED FOOD

MUCOSA LAYER
INNER STOMACH WALL

OBLIQUE MUSCLE LAYER
CIRCULAR MUSCLE LAYER
LONGITUDINAL MUSCLE LAYER

INNER SEROSA LAYER
OUTER SEROSA LAYER

LOOKING INTO THE STOMACH
Doctors can peer into the stomach using a flexible tube device called a gastroscope. The gastroscope is threaded down the throat and oesophagus, then passed through the cardiac sphincter to reach the stomach. It contains a light to illuminate the interior of the stomach, a fibre-optic system for viewing, and an air line for inflating the stomach to get a good view.

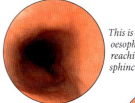

This is a view down the oesophagus, just before reaching the cardiac sphincter.

This is a view of the rugae, the folds lining the stomach.

Submucosa
This layer of loose, spongy tissue provides a cushion between the mucosa attached to its inner side and the muscle layers on its outer side. The submucosa also contains blood vessels and nerves.

Why you burp
Sometimes when you eat too quickly or drink fizzy drinks, you swallow air into your stomach along with your food. The air is pushed out of the stomach and up through the oesophagus as a burp.

Four-layered wall
Four main layers that make up the stomach wall – the red-brown mucosa, the submucosa, the muscularis or muscle layer, and the serosa – continue throughout most of the digestive tract. The flat, scale-like cells of the inner and outer serosa envelop the entire stomach.

Stomach lining
The velvety texture of the mucosa (inner wall) of the stomach is due to thousands of tiny dents or pits contained within it. Specialized groups of cells in the pits make and release hydrochloric acid. Other cells release enzymes such as pepsin, which softens tough meat fibres for easier digestion, and the hormone gastrin, which triggers the production of more gastric juices when food enters the stomach. As the stomach contracts, its mucosa is thrown into folds called rugae.

Barrier of mucus
What keeps the strong meat-digesting chemicals in the stomach from eating away at the stomach itself? The stomach's self defence is provided by a coat of slimy mucus. Groups of cells in the lining make a regular supply of mucus that lines and protects the interior.

ENOUGH TO LAST A LIFETIME
An average adult stomach processes about 500 kg (1,100 lb) of food each year. Over a 70-year lifespan, accounting for smaller meals during childhood, this adds up to more than 30,000 kg (66,000 lb) of food – the weight of six elephants without tusks!

The Liver, Pancreas, and Spleen

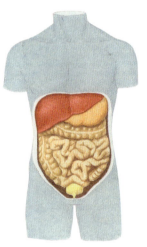

THE LIVER IS A complex chemical factory with at least 600 different roles in metabolism (body chemistry). It processes and stores body-building nutrients and energy-giving sugars; it filters your blood and recycles its constituents; it detoxifies dangerous chemicals; it produces bile for digestion; and it stores vitamins and minerals.

Under the left lobe of your liver is the pancreas, which has two distinct roles. One is to manufacture digestive chemicals known as pancreatic enzymes. These pour along a tube into your small intestine, where they help digest food. The other is to make insulin and glucagon, two hormones that control the way your body cells use energy. Also in your upper abdomen is the spleen, which helps defend against invading germs, as well as in filtering and maintaining healthy blood.

◁ Locating the liver

Like the stomach, the liver is higher in the body than many people realize. In fact, its uppermost hump lies just behind your right nipple, under the dome of the diaphragm. The pancreas is on your left side, just below your stomach. The spleen is tucked behind the left part of your stomach. All these organs are protected by your lower ribs.

▽ Inside a liver lobule

On the outside, the liver looks smooth and slightly rubbery. But it actually consists of around 75,000 tiny clusters called lobules. Each lobule, which is about 1 mm (0.04 in) across and shaped like a six-sided egg, has a central vein. Sheets of liver cells, or hepatocytes, fan out around it. Spaces between the cells, called sinusoids, are constantly topped up with oxygen-rich blood for the hard-working hepatocytes. Branches of the hepatic artery, portal vein, and hepatic duct surround each lobule.

Liver lobes
Weighing about 1.4 kg (3 lb), the liver is the body's largest internal organ. It consists of a large right lobe and a smaller left lobe, divided by the falciform ligament.

Gall bladder
Tucked into a dent under the right lobe of the liver is the gall bladder. This pear-shaped sac, about 10 cm (4 in) long, stores bile, a yellowish mixture of fluids, body salts, and wastes assembled by the liver. Bile helps to break up and digest fats in your food.

Double duty
The liver can "cover" for the gall bladder if it is removed, secreting bile directly into the small intestine at the duodenum. In fact, the liver itself is incredibly resilient. You could probably lose about three quarters of your liver, and it would still continue to function.

Bile tubes
The liver makes about a litre (almost two pints) of bile daily. Some trickles directly from the liver into the small intestine, along the hepatic duct and then the bile duct. The remainder is stored in the gall bladder. When the stomach squeezes food into the duodenum, a hormone is released to start the gall bladder's muscles squeezing, which squirts bile into the duodenum.

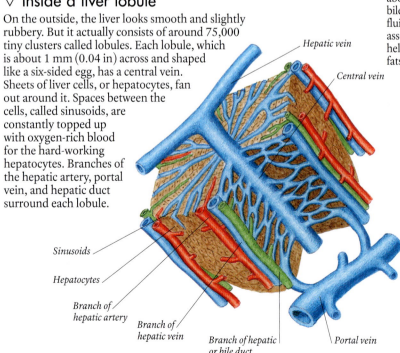

Hepatic vein

Central vein

Sinusoids

Hepatocytes

Branch of hepatic artery

Branch of hepatic vein

Branch of hepatic or bile duct

Portal vein

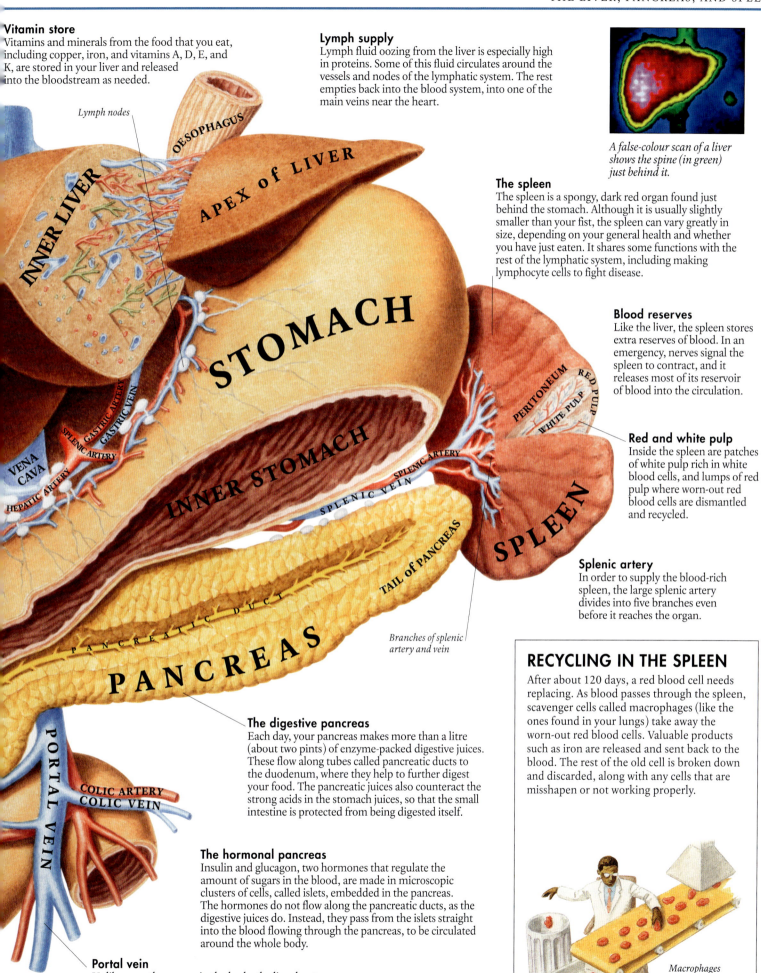

Vitamin store
Vitamins and minerals from the food that you eat, including copper, iron, and vitamins A, D, E, and K, are stored in your liver and released into the bloodstream as needed.

Lymph nodes

OESOPHAGUS

INNER LIVER

APEX of LIVER

Lymph supply
Lymph fluid oozing from the liver is especially high in proteins. Some of this fluid circulates around the vessels and nodes of the lymphatic system. The rest empties back into the blood system, into one of the main veins near the heart.

A false-colour scan of a liver shows the spine (in green) just behind it.

The spleen
The spleen is a spongy, dark red organ found just behind the stomach. Although it is usually slightly smaller than your fist, the spleen can vary greatly in size, depending on your general health and whether you have just eaten. It shares some functions with the rest of the lymphatic system, including making lymphocyte cells to fight disease.

Blood reserves
Like the liver, the spleen stores extra reserves of blood. In an emergency, nerves signal the spleen to contract, and it releases most of its reservoir of blood into the circulation.

STOMACH

GASTRIC ARTERY
GASTRIC VEIN
SPLENIC ARTERY
VENA CAVA
HEPATIC ARTERY

INNER STOMACH

SPLENIC VEIN
SPLENIC ARTERY

PERITONEUM
RED PULP
WHITE PULP

SPLEEN

Red and white pulp
Inside the spleen are patches of white pulp rich in white blood cells, and lumps of red pulp where worn-out red blood cells are dismantled and recycled.

TAIL of PANCREAS

Branches of splenic artery and vein

Splenic artery
In order to supply the blood-rich spleen, the large splenic artery divides into five branches even before it reaches the organ.

PANCREATIC DUCT

PANCREAS

PORTAL VEIN

COLIC ARTERY
COLIC VEIN

The digestive pancreas
Each day, your pancreas makes more than a litre (about two pints) of enzyme-packed digestive juices. These flow along tubes called pancreatic ducts to the duodenum, where they help to further digest your food. The pancreatic juices also counteract the strong acids in the stomach juices, so that the small intestine is protected from being digested itself.

The hormonal pancreas
Insulin and glucagon, two hormones that regulate the amount of sugars in the blood, are made in microscopic clusters of cells, called islets, embedded in the pancreas. The hormones do not flow along the pancreatic ducts, as the digestive juices do. Instead, they pass from the islets straight into the blood flowing through the pancreas, to be circulated around the whole body.

Portal vein
Unlike any other organ in the body, the liver has two blood supplies. It gets oxygen-rich blood from the hepatic artery, a branch of the aorta. But an additional supply of blood rich in nutrients comes from the stomach and intestines, through the portal vein.

RECYCLING IN THE SPLEEN
After about 120 days, a red blood cell needs replacing. As blood passes through the spleen, scavenger cells called macrophages (like the ones found in your lungs) take away the worn-out red blood cells. Valuable products such as iron are released and sent back to the blood. The rest of the old cell is broken down and discarded, along with any cells that are misshapen or not working properly.

Macrophages remove worn-out cells just as a worker discards faulty goods.

45

The Intestines

AFTER FOOD HAS BEEN GNASHED by your teeth, squeezed down your oesophagus, and pulped in your stomach, it enters the small and large intestines, where almost everything your body can use is extracted and absorbed.

Coiled intricately within your abdomen, the intestines form the main part of your digestive tract's length. Your small intestine is narrow, but long. It takes between one and six hours for food to travel through its 5–6 m (16–20 ft) length. The small intestine continues the chemical digestion of food that started in your mouth and stomach by bombarding it with more enzymes and digestive juices. Most of the resulting nutrients are absorbed into the blood and lymph vessels within its walls. At about 1.5 m (5 ft) long, your large intestine is shorter than the small intestine, but it is much wider. Its main roles are to absorb water and useful minerals from your food and prepare the leftovers for expulsion through the muscular ring that forms the final part of the tract, the anus.

Mesentery

The mesentery is part of the peritoneum, the thin membrane lining the inside wall of the abdomen. From its narrow end anchored to the rear abdominal wall, the mesentery fans out to wrap round the coils of the intestine. It carries blood and lymph to and from the intestine, and holds it in place so that it does not twist itself into knots.

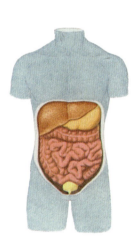

▷ Inside the digestive tubes

In this drawing of the intestines, the stomach and liver have been lifted away. The small intestine and the fatty pad known as the greater omentum are pulled to the left, stretching out the mesentery behind them. This reveals the arch of the large intestine, enriched with a web of blood vessels.

△ The body's guts

The small intestine folds and twists inside the centre of the abdomen, below the stomach, liver, and pancreas. The large intestine runs up the right side of the abdomen, across under the liver and stomach, and down the left side. It forms a "picture frame" around the small intestine, called the colonic loop.

An area of the intestinal wall about the size of your fingernail contains almost 3,000 villi.

Each villus is about 1 mm (0.04 in) long.

The ileum is the longest part of the small intestine.

Muscles contract and relax to squeeze food along.

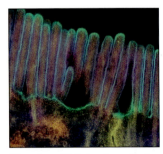

The microvilli lining the wall of the duodenum are shown in red in this false-colour micrograph, magnified about 20,000 times.

▷ Fingers for food

The inner lining of your small intestine is not smooth like a hose pipe, but folded and ridged, with a velvety texture. Within these folds are thousands of tiny, finger-shaped structures called villi. The outside of each villus is made of cells that are covered with hundreds of hair-like microvilli. Inside each villus is a dense network of small blood vessels, the arterioles and venules, and tiny lymph tubes known as lacteals. Nutrients from the food passing through the intestine seep through the thin covering of the microvilli into the blood and lymph tubes of the villi. There, they are carried away to nourish your body tissues. Together, the folds, ridges, villi, and microvilli increase the surface area of the small intestine by eight times, making digestion faster and more efficient.

Mucosa layer contains venules, arterioles, and thin muscle sheets.

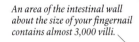

GREATER OMENTUM

JEJUNUM

MESENTERY

COILS of SMALL INTESTINE

ILEUM

SEROSA LAYER
MUSCLE LAYER
MUCOSA LAYER

INNER ILEUM

VENULE
LACTEAL
ARTERIOLE

VILLI

MICROVILLI

LACTEAL

VILLI

CIRCULAR MUSCLE LAYER

LONGITUDINAL MUSCLE

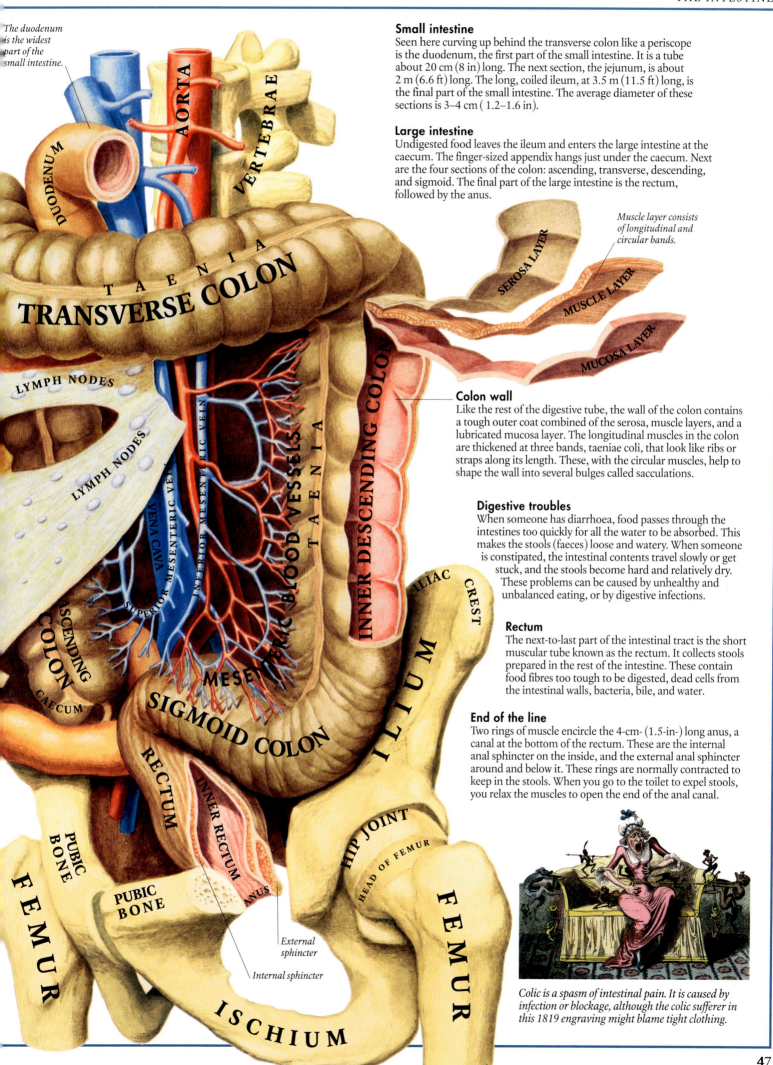

The duodenum is the widest part of the small intestine.

DUODENUM

AORTA

VERTEBRAE

TAENIA
TRANSVERSE COLON

LYMPH NODES

LYMPH NODES

VENA CAVA

SUPERIOR MESENTERIC VEIN

INFERIOR MESENTERIC VEIN

MESENTERIC BLOOD VESSELS

TAENIA

INNER DESCENDING COLON

ASCENDING COLON

CAECUM

MESENTERY

SIGMOID COLON

ILIUM

ILIAC CREST

RECTUM

INNER RECTUM

PUBIC BONE

PUBIC BONE

ANUS

HIP JOINT

HEAD OF FEMUR

FEMUR

FEMUR

ISCHIUM

External sphincter

Internal sphincter

Small intestine

Seen here curving up behind the transverse colon like a periscope is the duodenum, the first part of the small intestine. It is a tube about 20 cm (8 in) long. The next section, the jejunum, is about 2 m (6.6 ft) long. The long, coiled ileum, at 3.5 m (11.5 ft) long, is the final part of the small intestine. The average diameter of these sections is 3–4 cm (1.2–1.6 in).

Large intestine

Undigested food leaves the ileum and enters the large intestine at the caecum. The finger-sized appendix hangs just under the caecum. Next are the four sections of the colon: ascending, transverse, descending, and sigmoid. The final part of the large intestine is the rectum, followed by the anus.

Muscle layer consists of longitudinal and circular bands.

SEROSA LAYER

MUSCLE LAYER

MUCOSA LAYER

Colon wall

Like the rest of the digestive tube, the wall of the colon contains a tough outer coat combined of the serosa, muscle layers, and a lubricated mucosa layer. The longitudinal muscles in the colon are thickened at three bands, taeniae coli, that look like ribs or straps along its length. These, with the circular muscles, help to shape the wall into several bulges called sacculations.

Digestive troubles

When someone has diarrhoea, food passes through the intestines too quickly for all the water to be absorbed. This makes the stools (faeces) loose and watery. When someone is constipated, the intestinal contents travel slowly or get stuck, and the stools become hard and relatively dry. These problems can be caused by unhealthy and unbalanced eating, or by digestive infections.

Rectum

The next-to-last part of the intestinal tract is the short muscular tube known as the rectum. It collects stools prepared in the rest of the intestine. These contain food fibres too tough to be digested, dead cells from the intestinal walls, bacteria, bile, and water.

End of the line

Two rings of muscle encircle the 4-cm- (1.5-in-) long anus, a canal at the bottom of the rectum. These are the internal anal sphincter on the inside, and the external anal sphincter around and below it. These rings are normally contracted to keep in the stools. When you go to the toilet to expel stools, you relax the muscles to open the end of the anal canal.

Colic is a spasm of intestinal pain. It is caused by infection or blockage, although the colic sufferer in this 1819 engraving might blame tight clothing.

The Kidneys and Bladder

DOZENS OF WASTE PRODUCTS from chemical reactions build up inside your cells. These waste products seep into your blood, which carries them round your body to the kidneys, which filter the blood as it passes through. The kidneys extract the wastes, along with unwanted minerals and excess water, and prepare it for removal. The fluid wastes are called urine, which constantly trickles from the kidneys along two tubes known as ureters, to your bladder, a stretchable storage bag near the base of your abdomen. By the time it gets to the size of your clenched fist, sensors in the bladder wall are telling your brain that it's time to get rid of the urine. To do this, you relax a ring of muscle, the urethral sphincter, around the exit of your bladder. Muscles in the bladder wall squeeze the urine along a tube called the urethra, and out of your body.

▽ Position in the body

The two kidneys are higher in the body than many people realize. They are tucked into the rear upper abdomen, not far below the tips of the lungs, behind the lower parts of the liver and stomach, and shielded by the lowest ribs. They are not quite mirror images: the right kidney is usually 1–2 cm (0.4–0.8 in) lower than the left one.

Lung
Liver
Stomach
Kidney
Bladder

▽ Inside the kidney

About one million tiny filters, called nephrons, cleanse blood. As the blood flows through a nephron's microscopic knot of capillaries, the glomerulus, wastes and water are forced into a cup-shaped capsule around it. They then flow along a C-shaped tubule, where much of the water and useful minerals soak back into the blood. The remains trickle into the system of collecting tubes.

Your kidneys also help to balance the amount of water in your body. The water you lose every day, as sweat, urine, or vapour in your breath, needs to be replaced. If you do not get enough water from your meals and drinks, your kidneys reduce the amount of urine they produce. If you get too much water, they speed up production.

Renal blood vessels

Renal means to do with the kidneys. The wide, short renal arteries, one to each kidney, bring plenty of blood – over 1.2 litres (about a third of a gallon) each minute – to each kidney. This means that all the blood in your body passes through each kidney 400 times in one day. More than 99.9 per cent of this volume leaves the kidney along the renal vein. Less than 0.1 per cent is filtered out as urine.

Emergency glands

One suprarenal, or adrenal, gland sits at the top of each kidney. These glands make hormones that help us to act quickly when faced with an emergency.

Medulla

The inner kidney, or medulla, contains about 15 fan-like groups of collecting tubes for urine. These are separated by renal columns containing blood vessels.

A computer-coloured micrograph of a nephron shows the knot-like glomerulus (in blue), with the capsule (gold) around it.

Down the tubes

Urine passes through tiny holes called papillae at the bottom of the medulla, into a minor calyx. These tubes join to form larger ones, the major calyces, which eventually drain into an even bigger cavity called the pelvis.

Ureter

Each ureter is about 30 cm (12 in) long. For most of its length, it is as thick as the inner ink-holding tube of a ballpoint pen. But it is not a slack, hollow tube that urine rushes through, like water in a garden hose. Instead, the muscles in its thick walls contract in waves to massage urine downwards, to enter the bladder in a series of small spurts.

Diagram labels

SPLENIC AREA
COLIC AREA
KIDNEY
PANCREATIC AREA
JEJUNAL AREA
GASTRIC AREA
SUPRARENAL AREA

MEDULLA
PAPILLA
MEDULLA
RENAL COLUMN
MEDULLA
CORTEX OF KIDNEY
MINOR CALYX
MAJOR CALYX
BRANCH OF ARTERY
BRANCH OF VEIN
CORTEX
RENAL CAPSULE
FAT

SUPRARENAL GLAND
CAPSULE OF GLAND
MEDULLA OF GLAND
INFERIOR PHRENIC VEIN
PELVIS OF KIDNEY
URETER

RENAL ARTERY
RENAL VEIN
TESTICULAR ARTERY
TESTICULAR VEIN

Small vein
Small artery
U-shaped tubule in renal medulla.
Capillary network absorbs water and minerals.
Glomerulus and capsule in renal cortex.
Each kidney is wrapped in a layer of tissue called the renal capsule.
Urine collecting tube

MEDULLA
RENAL CORTEX
RENAL CAPSULE

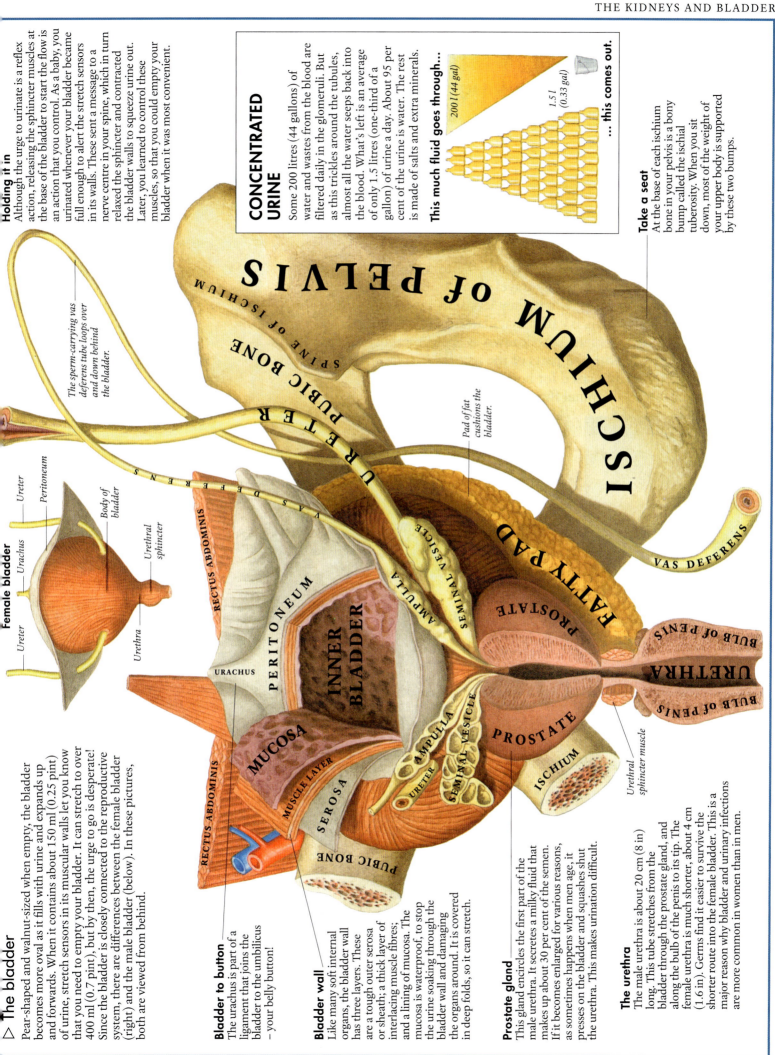

Holding it in

Although the urge to urinate is a reflex action, releasing the sphincter muscles at the base of the bladder to start the flow is an action that you control. As a baby, you urinated whenever your bladder became full enough to alert the stretch sensors in its walls. These sent a message to a nerve centre in your spine, which in turn relaxed the sphincter and contracted the bladder walls to squeeze urine out. Later, you learned to control these muscles, so that you could empty your bladder when it was most convenient.

CONCENTRATED URINE

Some 200 litres (44 gallons) of water and wastes from the blood are filtered daily in the glomeruli. But as this trickles around the tubules, almost all the water seeps back into the blood. What's left is an average of only 1.5 litres (one-third of a gallon) of urine a day. About 95 per cent of the urine is water. The rest is made of salts and extra minerals.

This much fluid goes through...

200 l (44 gal)

1.5 l (0.33 gal)

... this comes out.

Take a seat

At the base of each ischium bone in your pelvis is a bony bump called the ischial tuberosity. When you sit down, most of the weight of your upper body is supported by these two bumps.

ISCHIUM of PELVIS

SPINE OF ISCHIUM

PUBIC BONE

URETER

VAS DEFERENS

The sperm-carrying vas deferens tube loops over and down behind the bladder.

Pad of fat cushions the bladder.

FATTY PAD

PROSTATE

BULB OF PENIS

URETHRA

BULB OF PENIS

SEMINAL VESICLE

AMPULLA

VAS DEFERENS

RECTUS ABDOMINIS

PERITONEUM

INNER BLADDER

MUCOSA

MUSCLE LAYER

SEROSA

URACHUS

PUBIC BONE

AMPULLA

SEMINAL VESICLE

URETER

PROSTATE

ISCHIUM

Urethral sphincter muscle

Female bladder

Ureter

Urachus

Peritoneum

Ureter

Body of bladder

Urethral sphincter

Urethra

▷ The bladder

Pear-shaped and walnut-sized when empty, the bladder becomes more oval as it fills with urine and expands up and forwards. When it contains about 150 ml (0.25 pint) of urine, stretch sensors in its muscular walls let you know that you need to empty your bladder. It can stretch to over 400 ml (0.7 pint), but by then, the urge to go is desperate! Since the bladder is closely connected to the reproductive system, there are differences between the female bladder (right) and the male bladder (below). In these pictures, both are viewed from behind.

Bladder to button

The urachus is part of a ligament that joins the bladder to the umbilicus – your belly button!

Bladder wall

Like many soft internal organs, the bladder wall has three layers. These are a tough outer serosa or sheath; a thick layer of interlacing muscle fibres; and a lining of mucosa. The mucosa is waterproof, to stop the urine soaking through the bladder wall and damaging the organs around. It is covered in deep folds, so it can stretch.

Prostate gland

This gland encircles the first part of the male urethra. It secretes a milky fluid that makes up about 30 per cent of the semen. If it becomes enlarged for various reasons, as sometimes happens when men age, it presses on the bladder and squashes shut the urethra. This makes urination difficult.

The urethra

The male urethra is about 20 cm (8 in) long. This tube stretches from the bladder through the prostate gland, and along the bulb of the penis to its tip. The female urethra is much shorter, about 4 cm (1.6 in). Germs find it easier to survive the shorter route into the female bladder. This is a major reason why bladder and urinary infections are more common in women than in men.

The Male Reproductive System

THE SEXUAL ORGANS OF THE MALE BODY are specialized to produce and deliver sperm, the microscopic tadpole-shaped cells needed to fertilize the eggs produced within the female body. Sperm production begins at puberty and continues into old age. Sperm are produced within the testes. During sexual intercourse, they travel along a tube called the vas deferens to the penis, from which they leave the body. This is called ejaculation.

The reproductive organs, also known as the genital organs, are closely connected with the parts of the urinary system. In men, the urethra is the tube used for both the passage of sperm-containing semen (seminal fluid) and the flow of urine.

▷ Inside and out

The male reproductive organs are partly inside the lower abdomen, and partly hanging below it. Inside are glands such as the walnut-sized prostate, and tubes such as the vas deferens that link the organs. Outside are the penis and the bag of skin called the scrotum which hangs behind it. Suspended in the scrotum are the pair of egg-shaped testes, in which sperm cells are manufactured.

THE PATH OF SPERM

During sexual arousal and ejaculation, the sperm follow a path (shown in blue below) from their site of development in the testes, through the penis and out of the man's body. Along the way, several glandular organs contribute to the seminal fluid. These include two seminal vesicles behind the bladder, the prostate around the urethra, and two Cowper's, or bulbo-urethral, glands at the base of the penis. The fluid provides nutrients and chemicals to help the sperm on their way to fertilize an egg.

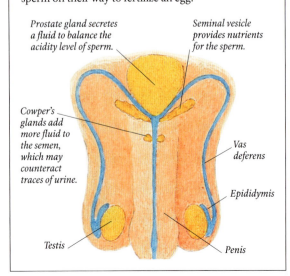

Prostate gland secretes a fluid to balance the acidity level of sperm.

Seminal vesicle provides nutrients for the sperm.

Cowper's glands add more fluid to the semen, which may counteract traces of urine.

Vas deferens

Epididymis

Testis

Penis

Vas deferens

The vas deferens is a 40-cm (16-in) tube that carries sperm from each testis and epididymis (a long coiled tube at the base of the vas deferens where sperm mature) to the penis. The vas deferens curls up and around the back of the bladder and down under it, towards the urethra. Just behind the base of the prostate, it joins the tube from the seminal vesicle and forms the ejaculatory duct, which then joins the main urethra.

Penis

The main part of the penis consists of three spongy cylinders, two corpora cavernosa (one of which can be seen above), and the corpus spongiosum. During sexual arousal, these fill up with blood, making the penis erect – that is, longer, thicker, and stiffer.

The loose flap of skin over the glans is called the foreskin.

ILIAC CREST of PELVIS

SKIN

TRANSVERSE ABDOMINAL

INTERNAL OBLIQUE MUSCLE

EXTERNAL OBLIQUE MUSCLE

ILIACUS MUSCLE

LATERAL CUTANEOUS NERVE of THIGH

FEMORAL NERVE

TESTICULAR VEIN AND ARTERY

DUCTUS (VAS) DEFEREN

FATTY LAYER

PUBIC BONE

SUSPENSORY LIGAMENT

RIGHT LEG

DORSAL ARTERY AND VEIN of PENIS

FATTY LAYER

SKIN

CORPUS CAVERNOSUM

CORPUS SPONGIOSUM

PENIS

SKIN

FORESKIN

GLANS PENIS

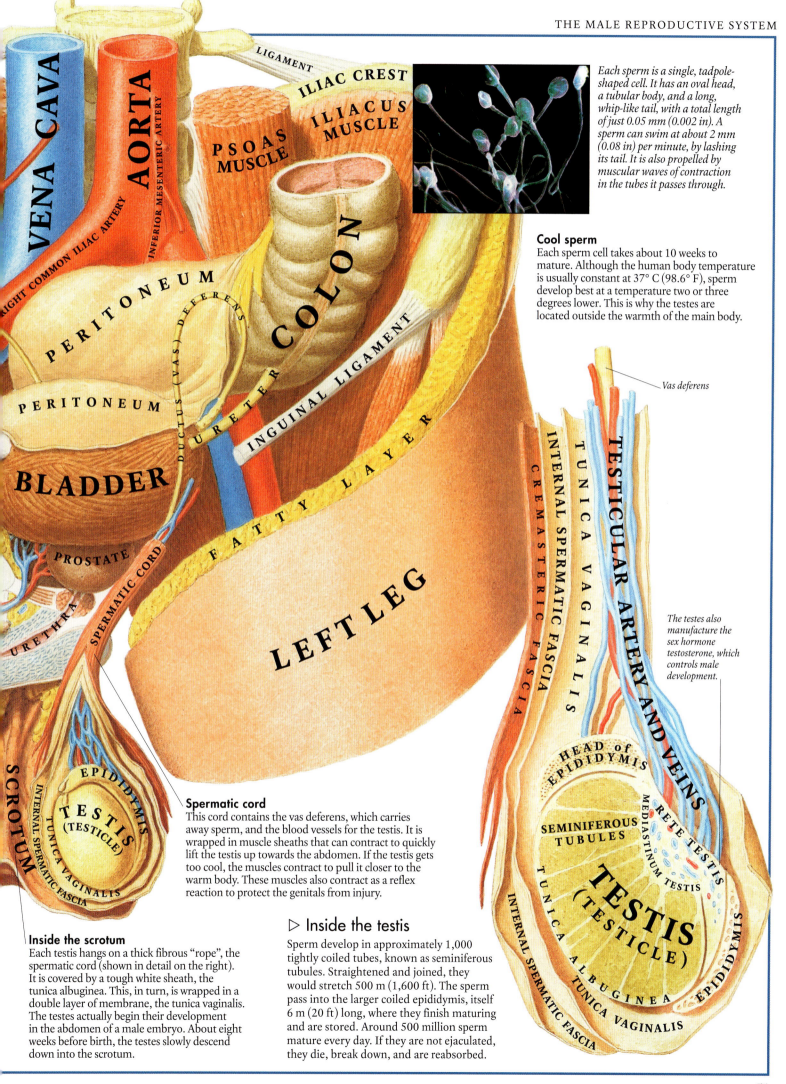

VENA CAVA

AORTA

RIGHT COMMON ILIAC ARTERY

INFERIOR MESENTERIC ARTERY

LIGAMENT

ILIAC CREST

PSOAS MUSCLE

ILIACUS MUSCLE

COLON

PERITONEUM

DUCTUS (VAS) DEFERENS

URETER

INGUINAL LIGAMENT

PERITONEUM

BLADDER

PROSTATE

FATTY LAYER

SPERMATIC CORD

LEFT LEG

URETHRA

SCROTUM

EPIDIDYMIS

INTERNAL SPERMATIC FASCIA

TESTIS (TESTICLE)

TUNICA VAGINALIS

CREMASTERIC FASCIA

INTERNAL SPERMATIC FASCIA

TUNICA VAGINALIS

TESTICULAR ARTERY AND VEINS

Vas deferens

The testes also manufacture the sex hormone testosterone, which controls male development.

HEAD of EPIDIDYMIS

SEMINIFEROUS TUBULES

RETE TESTIS

MEDIASTINUM TESTIS

TESTIS (TESTICLE)

TUNICA ALBUGINEA

INTERNAL SPERMATIC FASCIA

TUNICA VAGINALIS

EPIDIDYMIS

Each sperm is a single, tadpole-shaped cell. It has an oval head, a tubular body, and a long, whip-like tail, with a total length of just 0.05 mm (0.002 in). A sperm can swim at about 2 mm (0.08 in) per minute, by lashing its tail. It is also propelled by muscular waves of contraction in the tubes it passes through.

Cool sperm

Each sperm cell takes about 10 weeks to mature. Although the human body temperature is usually constant at 37° C (98.6° F), sperm develop best at a temperature two or three degrees lower. This is why the testes are located outside the warmth of the main body.

Spermatic cord

This cord contains the vas deferens, which carries away sperm, and the blood vessels for the testis. It is wrapped in muscle sheaths that can contract to quickly lift the testis up towards the abdomen. If the testis gets too cool, the muscles contract to pull it closer to the warm body. These muscles also contract as a reflex reaction to protect the genitals from injury.

▷ Inside the testis

Sperm develop in approximately 1,000 tightly coiled tubes, known as seminiferous tubules. Straightened and joined, they would stretch 500 m (1,600 ft). The sperm pass into the larger coiled epididymis, itself 6 m (20 ft) long, where they finish maturing and are stored. Around 500 million sperm mature every day. If they are not ejaculated, they die, break down, and are reabsorbed.

Inside the scrotum

Each testis hangs on a thick fibrous "rope", the spermatic cord (shown in detail on the right). It is covered by a tough white sheath, the tunica albuginea. This, in turn, is wrapped in a double layer of membrane, the tunica vaginalis. The testes actually begin their development in the abdomen of a male embryo. About eight weeks before birth, the testes slowly descend down into the scrotum.

The Female Reproductive System

EVERY FOUR WEEKS OR SO, depending on the timing of the menstrual cycle (explained opposite), an egg, or ovum, ripens in one of the two ovaries. This egg bursts from the ovary and is caught by the wispy fingers at the trumpet-shaped end of the fallopian tube. It is carried along the tube by a combination of massaging by the muscles in the tube wall, and waving movements of the tiny hair-like cilia in its lining. If sperm are present in the fallopian tube, this is the most common place for the egg to be fertilized. The egg continues its journey and emerges into the uterus. If it is fertilized, it embeds itself into the uterus, where it grows and develops into a baby. If it is not fertilized, it disintegrates and is expelled during menstruation.

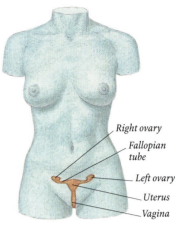

Right ovary
Fallopian tube
Left ovary
Uterus
Vagina

△ The reproductive organs

The main organs of the female reproductive system are the uterus, ovaries, and vagina. The uterus is about 7 cm (3 in) long and 5 cm (2 in) wide, unless it contains a baby.

▷ Inside the female abdomen

The reproductive organs nestle deep in the abdomen, beneath the intestines, in front of the lumbar (lower) vertebrae, and just behind the bladder. All of these organs are protected by the deep bowl formed by the bones of the surrounding pelvis, from the iliac crests at the sides to the pubic bones at the front.

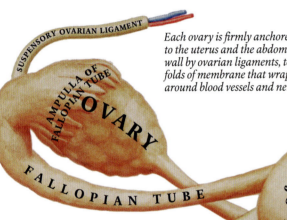

Each ovary is firmly anchored to the uterus and the abdominal wall by ovarian ligaments, tough folds of membrane that wrap around blood vessels and nerves.

The uterine ligament is one of several that secure the uterus to the sides of the lower abdomen.

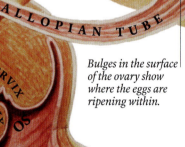

Bulges in the surface of the ovary show where the eggs are ripening within.

△ Ovaries and uterus

The uterus has a wide upper part called the body, which narrows to the cervix, or neck. The uterus meets the vagina at the os. The fallopian tubes, each about 10 cm (4 in) long, curve from the uterus to the ovaries. Each ovary is about the size of a large almond. Inside them, tiny eggs smaller than the dot on this "i" mature in sacs called follicles. About 600,000 immature eggs are present at birth, but only around 400 of these eventually mature. The ovaries also make the female hormones oestrogen and progesterone.

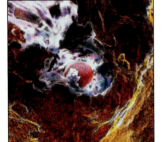

This computer-coloured scanning electron micrograph shows a ripe egg (in red) bursting from its follicle inside an ovary.

Folds in the inner mucosa layer of the vagina, called rugae, straighten out during intercourse or childbirth, as the vagina expands.

Under the mucosa is a layer of strong, interwoven muscle fibres.

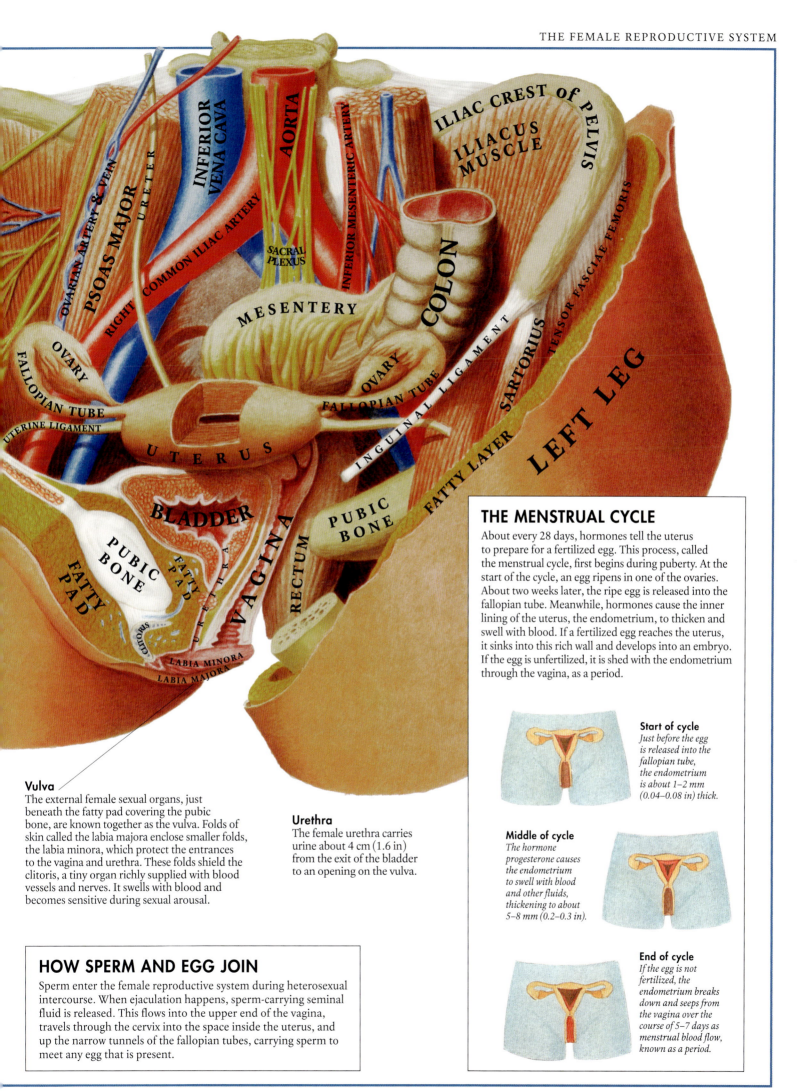

OVARIAN ARTERY & VEIN

PSOAS MAJOR

URETER

INFERIOR VENA CAVA

AORTA

RIGHT COMMON ILIAC ARTERY

SACRAL PLEXUS

INFERIOR MESENTERIC ARTERY

ILIAC CREST of PELVIS

ILIACUS MUSCLE

MESENTERY

COLON

TENSOR FASCIAE FEMORIS

OVARY

FALLOPIAN TUBE

UTERINE LIGAMENT

OVARY

FALLOPIAN TUBE

INGUINAL LIGAMENT

SARTORIUS

LEFT LEG

UTERUS

BLADDER

PUBIC BONE

FATTY PAD

FATTY PAD

URETHRA

CLITORIS

LABIA MINORA

LABIA MAJORA

VAGINA

RECTUM

PUBIC BONE

FATTY LAYER

THE MENSTRUAL CYCLE

About every 28 days, hormones tell the uterus to prepare for a fertilized egg. This process, called the menstrual cycle, first begins during puberty. At the start of the cycle, an egg ripens in one of the ovaries. About two weeks later, the ripe egg is released into the fallopian tube. Meanwhile, hormones cause the inner lining of the uterus, the endometrium, to thicken and swell with blood. If a fertilized egg reaches the uterus, it sinks into this rich wall and develops into an embryo. If the egg is unfertilized, it is shed with the endometrium through the vagina, as a period.

Start of cycle
Just before the egg is released into the fallopian tube, the endometrium is about 1–2 mm (0.04–0.08 in) thick.

Middle of cycle
The hormone progesterone causes the endometrium to swell with blood and other fluids, thickening to about 5–8 mm (0.2–0.3 in).

End of cycle
If the egg is not fertilized, the endometrium breaks down and seeps from the vagina over the course of 5–7 days as menstrual blood flow, known as a period.

Vulva
The external female sexual organs, just beneath the fatty pad covering the pubic bone, are known together as the vulva. Folds of skin called the labia majora enclose smaller folds, the labia minora, which protect the entrances to the vagina and urethra. These folds shield the clitoris, a tiny organ richly supplied with blood vessels and nerves. It swells with blood and becomes sensitive during sexual arousal.

Urethra
The female urethra carries urine about 4 cm (1.6 in) from the exit of the bladder to an opening on the vulva.

HOW SPERM AND EGG JOIN

Sperm enter the female reproductive system during heterosexual intercourse. When ejaculation happens, sperm-carrying seminal fluid is released. This flows into the upper end of the vagina, travels through the cervix into the space inside the uterus, and up the narrow tunnels of the fallopian tubes, carrying sperm to meet any egg that is present.

The Development of a Baby

A BABY STARTS TO DEVELOP when a sperm cell from the father joins and fertilizes an egg cell from the mother. The resulting single cell immediately begins a series of cell divisions that will gradually shape a tiny human body. About a week after fertilization, the growing cluster of cells embeds itself in the lining of the uterus. During the next six or seven days, the hollow ball of cells divides into two main parts. The embryo develops in one section. After a time, it will receive nourishment from the mother's blood via the other section, which develops into the placenta and umbilical cord.

In general, the baby's development is "head first". The spinal column and brain appear very early, then the other parts of the head, the heart, and vital organs in the torso, followed by the arms and legs. About two months after fertilization, the baby is as big as your thumb – yet it has all its main body parts.

STAGES OF PREGNANCY

Fertilization
This microphotograph shows the head of a sperm cell, the pink "ball" lower right, about to merge into the relatively huge egg cell. Several hours later, the fertilized egg divides into two cells. These divide, in turn, and so on. After five days, the cells form a hollow ball called the blastocyst which floats free in the uterus.

Wall of uterus

Implantation
About 7–10 days after fertilization, the outside layer of the pinpoint-sized blastocyst has broken down and it has begun to "burrow" its way into the wall of the uterus. This stage is called implantation. The uterine lining is richly supplied with blood vessels, so the blastocyst is surrounded by nourishment. Arched layers of cells inside the ball will become the baby.

These cells form the placenta.

These cells form the embryo.

Brain

Embryonic growth
Cells continue to multiply and move, and begin to develop into different types and shapes, such as nerve cells and blood cells. About three weeks after fertilization, two large bulges mark the growing brain. In the eight weeks after fertilization, the developing baby is called an embryo.

Spinal cord

Wall of uterus

Yolk sac provides nutrients

Amniotic fluid

The nutrients in the yolk sac are almost used up. It will shrink and wither away.

Developing placenta

The body stalk connects to the remains of the yolk sac. The umbilical cord is now developing.

From embryo to foetus
Two months after fertilization, huge changes have taken place. The baby is recognizably human, with a face, eyes, ears, and mouth, and all its major organs. It floats in amniotic fluid, warm and cushioned from bumps. From this time until birth it is called a foetus.

Between the eighth week and birth, the length of the foetus will increase more than 20 times.

◁ The placenta
This organ is about the size and shape of a dinner plate – which is apt, since it passes food from mother to baby. Used blood (shown here in blue), low in oxygen and nutrients, flows from the baby's heart along two umbilical arteries to the placenta. Here it absorbs oxygen and nutrients, and gets rid of wastes. Refreshed and "red" again, the blood flows back to the baby along the umbilical vein.

The placenta makes hormones that help to control pregnancy and birth.

Harmful substances, such as alcohol or drugs, can pass from the mother's blood to the foetus.

Pool of mother's blood

The baby's blood passes through tiny vessels that run through pools of the mother's blood. The blood supplies do not actually mix.

The umbilical cord is about 50 cm (20 in) long and 1–2 cm (0.4–0.8 in) thick. The blood vessels are embedded in a tough, jelly-like substance that prevents kinks and tight knots.

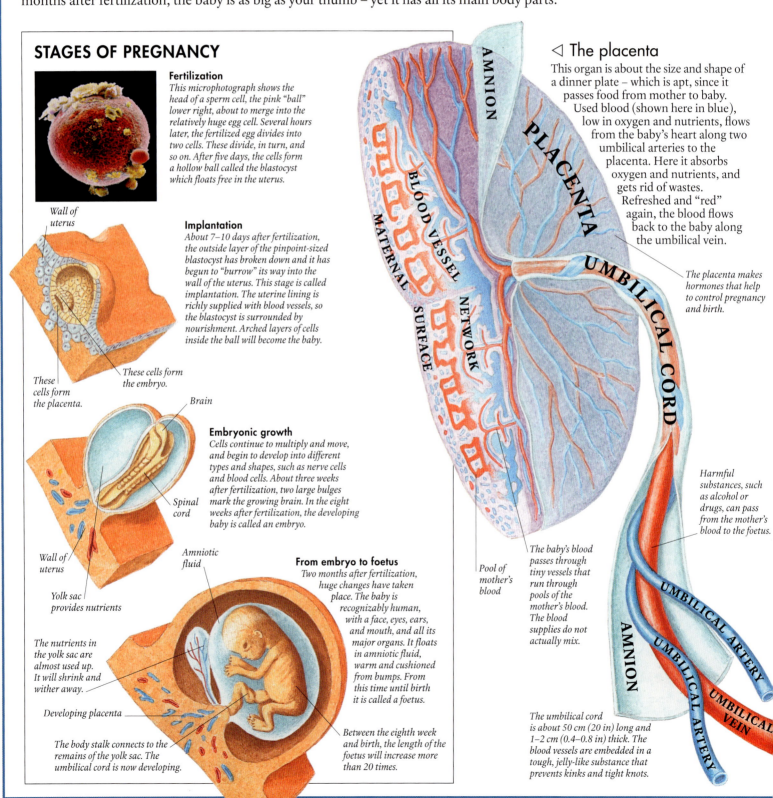

AMNION

PLACENTA

BLOOD VESSEL

MATERNAL SURFACE

NETWORK

UMBILICAL CORD

AMNION

UMBILICAL ARTERY

UMBILICAL VEIN

UMBILICAL ARTERY

▽ Waiting for birth

Nine months after fertilization, this fully developed baby is waiting to be born. It still has the lifeline of the umbilical cord and receives oxygen and nutrients from the placenta. Within seconds of birth, its hearty cries will open up its own lungs and drain the amniotic fluid from them, so that it can take its first breaths of air.

Changes in the uterus

The muscular wall of the uterus stretches enormously as the baby grows. It is now the largest muscle in the mother's body. It contracts strongly and periodically at birth, to push the baby through the cervix and vagina, out into the world.

Foetal membranes and fluid

The baby is wrapped in two thin, semi-transparent "bags" or membranes. The inner one is called the amnion and contains about 1 litre (1.76 pints) of amniotic fluid, which the baby swallows and into which it urinates. The fluid contains small quantities of proteins, fats, sugars, minerals, hormones, and enzymes. The outer membrane, the chorion, comes from the same cells that formed the placenta.

Expanding uterus squeezes organs above

The site of the placenta varies. In this pregnancy, it has formed at the front of the uterus.

By the seventh month the foetus is fully developed and, if necessary, could survive outside the uterus. In the last two months the foetus grows in size and puts on weight.

WALL OF UTERUS

INFERIOR VENA CAVA

AORTA

PSOAS MUSCLE

ILIAC CREST OF PELVIS

ILIACUS MUSCLE

UMBILICAL CORD

CHORION

AMNION

PLACENTA

Myometrium (muscle layer) of uterus

ABDOMINAL WALL

FATTY LAYER

BLADDER

PUBIC BONE

CERVIX

URETHRA

VAGINA

CLITORIS

FATTY PAD

RIGHT LEG

LEFT LEG

Squashed flat

Pregnancy brings enormous changes in the expectant mother's abdomen – turn to the previous page and see the difference! The baby takes up so much space that organs other than the uterus are squashed flat. The mother may need to urinate more frequently because her bladder does not have as much room to expand.

Twist and turn

A baby is active inside the uterus, changing its position, kicking its legs, and waving its arms about. The mother can feel these movements from about four months into the pregnancy. The baby also spends part of its day sleeping.

Exit from the uterus

During pregnancy, the cervix (neck of the uterus) is tightly closed and plugged with mucus. As birth approaches, the muscles of the cervix relax and the mucus plug slips out, so that the baby's head and body can pass through. Contractions of the muscles of the uterus push the baby's head down as the cervix widens. The process of giving birth is called "labour". When the baby is born, the umbilical cord may be clamped and cut.

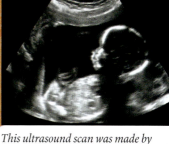

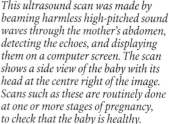

This ultrasound scan was made by beaming harmless high-pitched sound waves through the mother's abdomen, detecting the echoes, and displaying them on a computer screen. The scan shows a side view of the baby with its head at the centre right of the image. Scans such as these are routinely done at one or more stages of pregnancy, to check that the baby is healthy.

The Lower Back

SPRAINED JOINTS, strained muscles, and slipped discs – lower back problems seem to affect nearly everyone from time to time. This may be partly due to our upright posture, which is unique among our close relatives, the apes and monkeys. We keep our spines almost vertical as we walk and stand, which puts a great deal of strain on the lower back. This region must support the weight of the torso, head, neck, and arms. It is especially stressed when we lean forward, twist, or bend down to lift heavy objects.

◁ The weight of the world

In the mythical stories of the ancient Greeks, Atlas was one of the Titans who lost a war against the chief god, Zeus, and his Olympians. As a punishment, Atlas had to hold up the sky for ever. He is often pictured as a well-muscled man, the strength in his upper and lower back bearing the weight of the planet on his shoulders. The topmost bone of the spine, the atlas, is named after him because it supports the "globe" of the skull. Books of maps are also called atlases because a collection of maps published in the 16th century featured a drawing of Atlas on its cover.

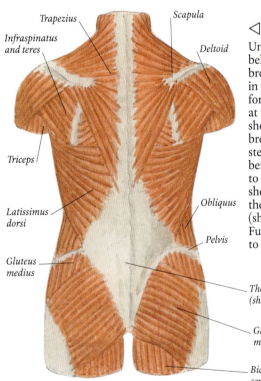

Trapezius

Infraspinatus and teres

Scapula

Deltoid

Triceps

Latissimus dorsi

Gluteus medius

Obliquus

Pelvis

Thoracolumbar fascia (sheet of fibrous tissues)

Gluteus maximus

Biceps femoris and semitendinosus

◁ Band of back muscles

Under the skin and the fatty layer just below it, the back is criss-crossed by broad bands of muscle. The muscles in your lower back provide support for your upright posture. The muscles at the top of the back move your shoulders and arms, and help you to breathe. Those in the central group steady the spine, and enable you to bend forwards and back, and twist to the side. Overlapping muscle sheets join this central region to the bony projections on the scapulae (shoulder blades) and pelvis (hips). Further muscles link these bones to the arms and legs.

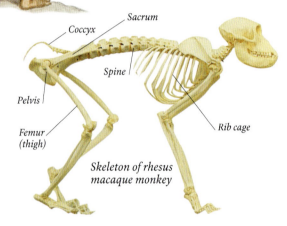

Coccyx

Sacrum

Spine

Pelvis

Femur (thigh)

Rib cage

Skeleton of rhesus macaque monkey

△ Four-legged relative

This monkey, like most of our mammal relations, moves about on all fours. Its spine and thighs are approximately at right angles to each other. But this arrangement could not support a human's two-legged gait. During 5–10 million years of evolution, the human lower spine, hip bones, and hip joints have tilted so that the legs are directly beneath the main body, to support its balanced weight. This not only keeps us from toppling over, but also frees our arms and hands for grasping and manipulation.

THE PELVIC BONES

The bowl-shaped ring of bones at the base of the lower torso is called the pelvis. It is formed by two innominate or hip bones at each side that curve round to meet at the pubic symphysis at the front, with the triangular base of the spine, the sacrum, between them at the back. Each hip bone consists of three individual bones that fuse together during childhood: the flank bone, or ilium, the gluteal bone, or ischium, and the pubic bone, or pubis. A woman's pelvis has a larger and more rounded hole in the centre than a man's. This makes it easier for a baby to pass through this hole, known as the birth canal, as it leaves the uterus at birth.

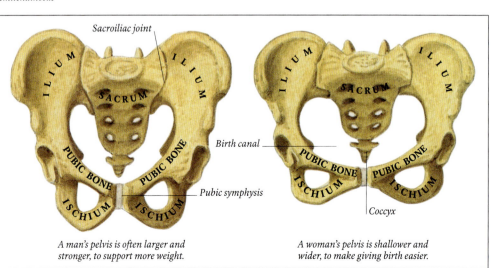

Sacroiliac joint

ILIUM

SACRUM

PUBIC BONE

ISCHIUM

Birth canal

Pubic symphysis

Coccyx

A man's pelvis is often larger and stronger, to support more weight.

A woman's pelvis is shallower and wider, to make giving birth easier.

▽ At the base of the back

In this view of the lower torso from the rear, the main muscles and tough sheaths of fascia beneath the skin are shown on the right side. They are omitted from the left side, revealing the vertebrae of the spine with its network of spinal nerves, and the hip bones joined to the sacrum at its base. The left kidney and the ureter can be seen resting against the rear surfaces of the intestines in front of them.

Around the back

The iliocostalis lumborum has several straps that link the lumbar vertebrae and sacrum with the lower six ribs. It is part of the long, complicated, muscle-and-tendon group, the erector spinae.

Bending over backwards

The erector spinae group runs down the side of the back, from the base of the skull to the hips and buttocks. The seven muscles in this group, including the longest, the longissimus thoracis, allow you to twist, lean, arch your back, and bend over backwards.

Inferior vena cava

Transverse colon rests at the top of the coils of the small intestine.

Tendons attach straps of iliocostalis to the ribs.

Oblique muscles twist the torso.

You can feel this crest through your skin.

FATTY LAYER

SKIN

EXTERNAL OBLIQUE

INTERNAL OBLIQUE

DESCENDING COLON

TAENIA

LEFT KIDNEY

RENAL ARTERY

RENAL VEIN

URETER

ABDOMINAL AORTA

FIRST

SECOND

LUMBAR NERVE PLEXUS

THIRD

LUMBAR (BACK) NERVE

FOURTH

FIFTH LUMBAR VERTEBRA

LONGISSIMUS THORACIS

ILIOCOSTALIS LUMBORUM

TENTH RIB

LATISSIMUS DORSI

EXTERNAL OBLIQUE

ILIAC CREST of PELVIS

ILIUM

PELVIS

PERONEAL NERVE

TIBIAL NERVE

SACRUM

ISCHIUM

RECTUM

Coccyx

THORACOLUMBAR FASCIA

ILIAC CREST

GLUTEUS MEDIUS

GLUTEUS MAXIMUS

PELVIS

Nerve plexus

A plexus is an interwoven network of nerves (or blood vessels). On either side of the lower back, the nerves running to and from the organs in the pelvic area, and to the muscles in the hip, thigh, and leg, form the lumbar and sacral plexuses.

Sacral plexus

Iliac artery

Iliac vein

Into the leg

The pelvic area is one of the body's major junctions. Blood vessels, nerves, and lymph vessels branch out here to supply the structures within the abdomen. Then they divide to send branches into each leg. The iliac arteries are divisions of the abdominal aorta. The iliac veins meet to form the inferior vena cava.

Sacrum

The forces generated by your upper body and your lower body meet at the sacrum, the strong, triangular wedge of five fused bones that is the only connection between the spinal column and the pelvis.

Gluteus medius

This muscle keeps your torso upright when one foot is on the ground and the other is off, as in walking or running.

Muscular fascia

Fascias are sheets, tubes, straps, ribbons, and bands of fibrous material that represent the body's "wrapping paper". Compared to most body parts they have a sparse blood supply, hence their pale colour. Fascias cover many organs, and pack the spaces between them. Some individual muscles and muscle groups are encased in fascial sheaths. These keep the muscles together and help to anchor them to nearby bones, as in the thoracolumbar fascia above the crest of the pelvis.

The Leg and Foot

WALKING AND STANDING might seem easy, natural movements, but watch a toddler totter with unsteady steps, or wobble while learning to stand. You soon realize what an effort these processes can be! Our two-legged gait is naturally unstable. Stand in one place for an hour, and your muscles begin to ache from the strain of the constant adjustments they make to keep your body balanced over your feet.

Before humans evolved to walk upright, arms and legs were very similar. The arrangements of bones and muscles is still much the same. The hip corresponds with the shoulder, and the knee with the elbow. But over millions of years our legs have become adapted to carry the weight of our body, while our arms are more suited to flexibility.

▽ Peg-leg sailor

In days gone by, sailors risked "life and limb". Many an old sea-dog lost a leg – to the jaws of a shark, smashed by falling timbers, or caught in a fast-tightening rope. The leg contains no vital organs, but it receives a copious blood supply to its powerful muscles. So after the injury, the first priority would be to staunch the wound. As the stump healed, the next job would be to carve a replacement.

One famous "peg-leg" was Captain Ahab, in Herman Melville's book Moby Dick.

▽ Down the leg

In this front view of the right leg, some muscles have been moved to reveal the various arteries, veins, and nerves snaking below and between them. In general, the front leg muscles straighten the knee, bend the ankle, and curl the toes. The muscles at the rear of the leg, shown below right, bend the knee and straighten the ankle.

Knee-kickers

Four large muscles cover the front and sides of the thigh. They are the rectus femoris (rectus means "straight"), and the three-part vastus: the vastus lateralis, medialis, and intermedius. Together they make up the quadriceps femoris muscle group, and they are used to straighten a bent knee, for example, when kicking a football.

A hinge for your leg

Your body's largest joint is the knee joint, chiefly between the femur and tibia. It works like the hinge joint in your elbow, so that you can fold your legs under when you kneel or stretch them out. It can only swivel slightly, helping you to turn your foot to point your toes out or in.

Hurdling athletes must swing two 10-kg (22-lb) weights – their legs – to and fro several times each second, and lift them over the hurdles.

The iliac artery branches out to carry fresh blood to the entire leg.

ILIAC ARTERY

PUBIC BONE

INGUINAL LIGAMENT

PSOAS
ILIACUS MUSCLE

ILIUM

TENSOR FASCIA LATA

SKIN OF THIGH

SARTORIUS

FEMORAL NERVE

FEMORAL ARTERY

FEMORAL VEIN

GREAT SAPHENOUS VEIN

GRACILIS

ADDUCTOR LONGUS

SARTORIUS

RECTUS FEMORIS

VASTUS LATERALIS

VASTUS MEDIALIS

FATTY LAYER

QUADRICEPS TENDON

SKIN

PERONEAL NERVE

There are about 24 muscles in the hip and thigh region, which are used to swing the thigh to the side, back, and front.

Knees up

The gracilis brings the knee up and pulls it across the front, towards the middle of the body.

Longest muscle

The body's longest muscle is the sartorius. It runs like a diagonal belt down from the crest of the hip bone, across the front of the thigh, and along the inner side of the knee, where it is anchored on the upper end of the tibia. It helps to bend both hip and knee, and to twist the leg. This muscle allows you to sit cross-legged.

Wobbly knees

The patella, or kneecap, forms an odd sliding joint with the lower front of the femur. This disc-shaped bone lies within the shared tendon of the quadriceps femoris muscle group. If you sit on the floor with your leg stretched in front of you, and relax your leg muscles, you can move the patella gently from side to side with your hand.

The sartorius muscle has been cut away here to show the muscles in the front of the thigh.

Labels on skeleton diagram:
CREST OF ILIUM
ILIUM
ISCHIUM
Knee joint
FEMUR
TIBIA
Hip joint
Popliteal (knee) fascia
Medial head
Fibula (calf bone)
Tarsals (ankle bones)

Labels on muscle diagram:
CREST OF ILIUM
GLUTEUS MAXIMUS
SEMIMEMBRANOSUS
GRACILIS
SEMITENDINOSUS
BICEPS FEMORIS
GASTROCNEMIUS
GASTROCNEMIUS
ACHILLES TENDON
SOLEUS
Lateral head

▽ A rear view

Like the front of the leg, the rear is swathed in bulky muscles. They form bulges that can be felt through the layers of skin and fat. The main calf muscle, the gastrocnemius, has two major parts, or heads. They are named using two common anatomical terms: medial (toward the middle) and lateral (to the side). The medial head is nearer the middle of the body, and the lateral head is on the outer side.

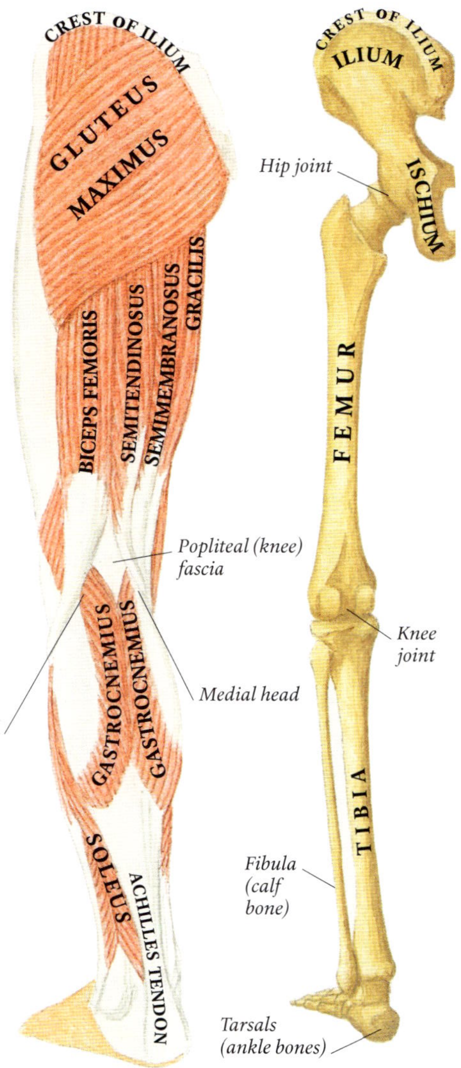

THE MECHANICS OF WALKING

Walking has been called "controlled falling". You start by tipping your head and torso forward, to the point of overbalance. Then you extend one leg to stop yourself toppling over – and repeat the process. During the power stride, your body weight transfers from your heel to your toe, as you push off and step ahead. Your pelvis tilts and the weight shifts to your other foot during the recovery stride.

As you step along, your leg muscles contract and relax in split-second unison. The muscles in your torso, arms, neck and head are also hard at work. The main thrust for each step is provided by the muscles in the buttock and rear thigh, pulling your femur (thigh bone) down and back.

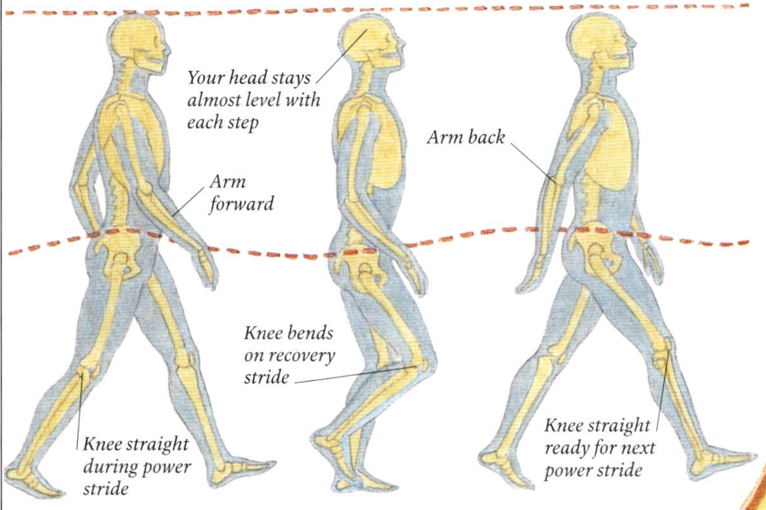

Diagram labels (walking figures):
Your head stays almost level with each step.
Arm forward
Arm back
Knee bends on recovery stride
Knee straight during power stride
As one leg goes forward, you extend the opposite arm, for balance.
Your pelvis swings between each step.
Knee straight ready for next power stride
Swinging your arms and legs gives you forward momentum.

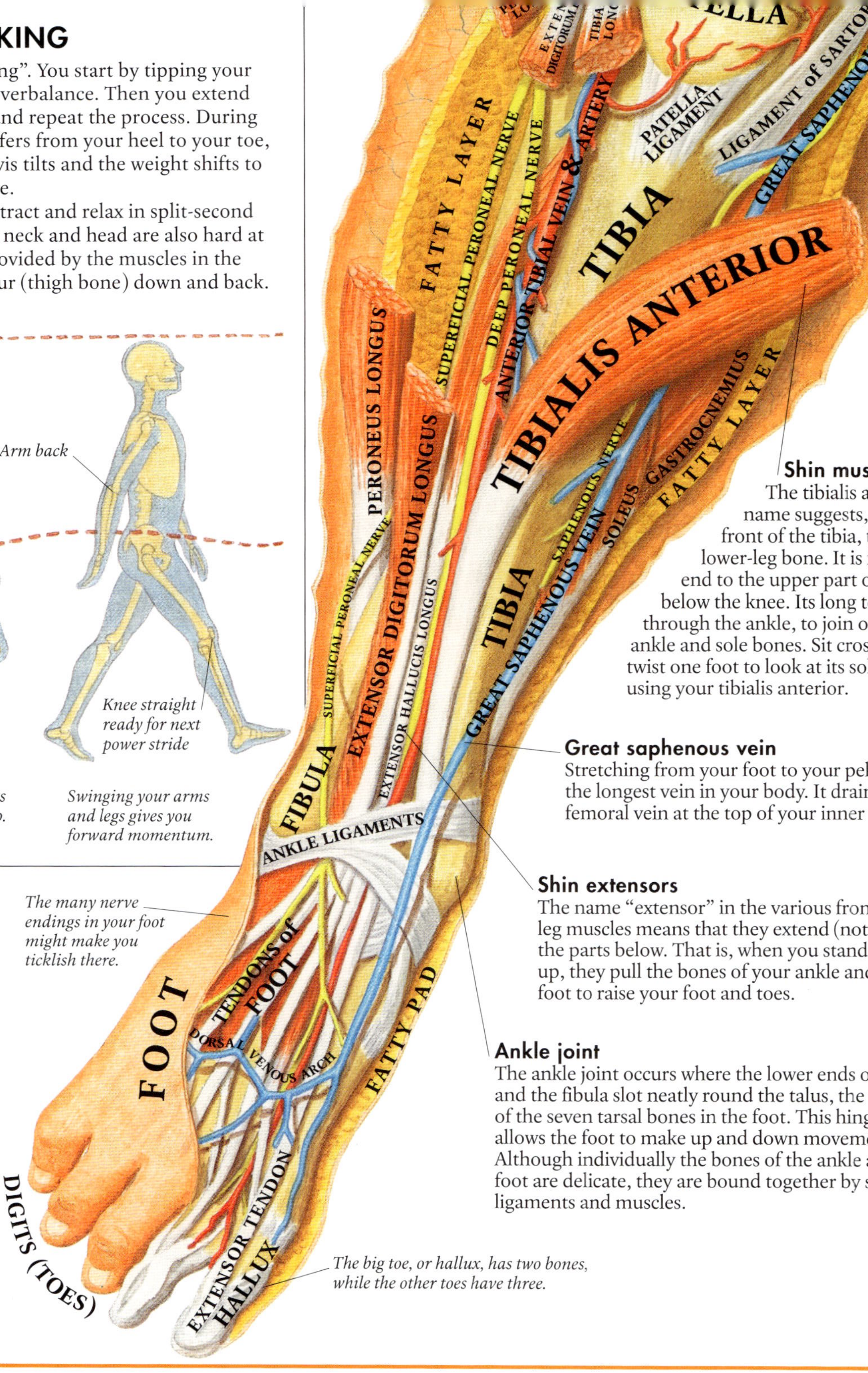

▷ The spring in your step

The foot lacks the delicate dexterity of the hand. Instead it is a flexible platform, whose elastic arched construction absorbs bumps and hollows in the ground as you walk, and puts the spring in your step. The longest nerves in the body connect your toes with your brain far above.

Toe bones
The bones of the digits, or toes, help you to balance when standing, and give the foot the grip, and some of the strength, to push off the ground when walking or running.

Labels on leg/foot dissection:
PATELLA
LIGAMENT OF SARTORIUS
GREAT SAPHENOUS VEIN
PATELLAR LIGAMENT
TIBIALIS ANTERIOR
GASTROCNEMIUS
TIBIA
FATTY LAYER
TIBIALIS ANTERIOR
SOLEUS
SAPHENOUS NERVE
GREAT SAPHENOUS VEIN
TIBIA
EXTENSOR DIGITORUM LONGUS
ANTERIOR TIBIAL VEIN & ARTERY
DEEP PERONEAL NERVE
SUPERFICIAL PERONEAL NERVE
FATTY LAYER
EXTENSOR HALLUCIS LONGUS
PERONEUS LONGUS
SUPERFICIAL PERONEAL NERVE
FIBULA
ANKLE LIGAMENTS
FATTY PAD
DORSAL VENOUS ARCH
TENDONS OF FOOT
FOOT
EXTENSOR HALLUX
DIGITS (TOES)

Shin muscle
The tibialis anterior, as its name suggests, sits on the front of the tibia, the main lower-leg bone. It is fixed at its top end to the upper part of the tibia, just below the knee. Its long tendon goes through the ankle, to join onto the inner ankle and sole bones. Sit cross-legged, and twist one foot to look at its sole. You are using your tibialis anterior.

Great saphenous vein
Stretching from your foot to your pelvis, this is the longest vein in your body. It drains into the femoral vein at the top of your inner thigh.

Shin extensors
The name "extensor" in the various front lower leg muscles means that they extend (not flex) the parts below. That is, when you stand up, they pull the bones of your ankle and foot to raise your foot and toes.

Ankle joint
The ankle joint occurs where the lower ends of the tibia and the fibula slot neatly round the talus, the topmost of the seven tarsal bones in the foot. This hinged joint allows the foot to make up and down movements. Although individually the bones of the ankle and foot are delicate, they are bound together by strong ligaments and muscles.

The big toe, or hallux, has two bones, while the other toes have three.

The many nerve endings in your foot might make you ticklish there.

59

The Hip and Knee

YOUR HIP IS A ball-and-socket joint. The rounded head of your biggest bone, the femur (thigh bone), fits snugly into a cup-shaped socket on your pelvis (hip bone). The hip is your leg's equivalent of your arm's shoulder. It is a much stronger joint, bound by tough ligaments and surrounded by your body's most powerful muscles. It is also more stable, to stand the stresses of walking. What you gain in stability, however, you lose in mobility. You can swing your leg a fair way forward, less to the side, and only a little to the rear. Big movements, such as drawing your leg back, ready to take a kick, are only possible because the whole pelvis tilts over the other hip.

The knee joint, shown in detail on the opposite page, works like a hinge to move your shin and foot forwards and backwards. Unlike the hinge joint in your elbow, it is a meeting of only two bones. Two knuckle-like bumps on the lower end of the femur sit on twin dents in the upper part of the tibia.

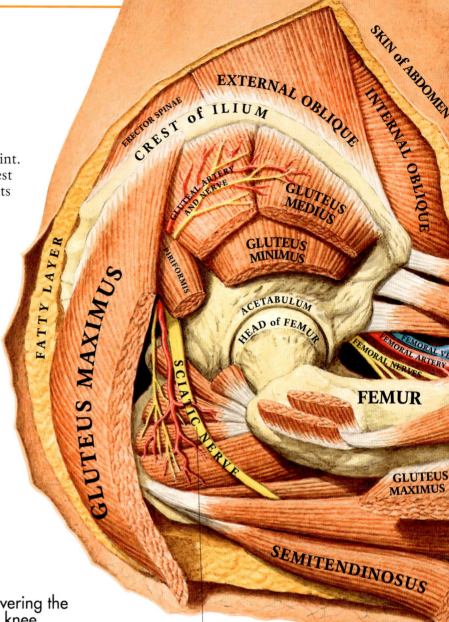

SKIN of ABDOMEN
INTERNAL OBLIQUE
EXTERNAL OBLIQUE
ERECTOR SPINAE
CREST of ILIUM
GLUTEAL ARTERY AND NERVE
GLUTEUS MEDIUS
GLUTEUS MINIMUS
PIRIFORMIS
FATTY LAYER
ACETABULUM
HEAD of FEMUR
FEMORAL VE
FEMORAL ARTERY
FEMORAL NERVES
FEMUR
GLUTEUS MAXIMUS
SCIATIC NERVE
GLUTEUS MAXIMUS
SEMITENDINOSUS

A skier bends the body at the knees, hips, and ankles, to stay stable and in control when skiing over bumps and hollows in the snow below.

▷ Uncovering the hip and knee

The picture on the right delves deep into the leg, under the skin and many muscles, to uncover the details of the hip and knee joints. The socket of the pelvis, slightly smaller than your cupped hand, is at the junction of all three hip bones on each side, the ilium, ischium, and pubis. The femur in the thigh is your longest and strongest bone, representing about one-quarter of your total height.

Sciatic nerve

Stretching from the sacrum in the spine all the way down through the hamstring muscles in the thigh, this is the longest nerve in your body. The sciatic nerve carries messages to and from all of the muscles of the leg and foot, as well as supplying the back of the thigh. It divides into the common peroneal and tibial nerves at around knee level.

FEEL A JERK

A reflex is an automatic reaction which your body does "on its own", without your brain thinking about it. You have dozens of reflexes, from blinking your eyelid shut when something gets too close to your eye, to sneezing when something gets up your nose. Doctors test your knee-jerk reflex to make sure your nerves are working well. A tap just below the kneecap stretches a tendon of the front thigh muscle. Sensors detect its movement and send nerve signals to the spinal cord. Reflex nerve connections in the cord send signals straight back out again, without "telling" your brain. These cause the thigh muscles to contract, jerking up your shin. This reflex may help you to keep upright after a long time on your feet.

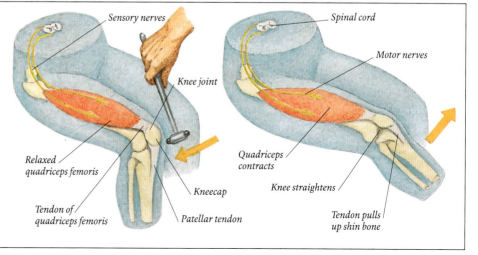

Sensory nerves
Knee joint
Relaxed quadriceps femoris
Tendon of quadriceps femoris
Kneecap
Patellar tendon

Spinal cord
Motor nerves
Quadriceps contracts
Knee straightens
Tendon pulls up shin bone

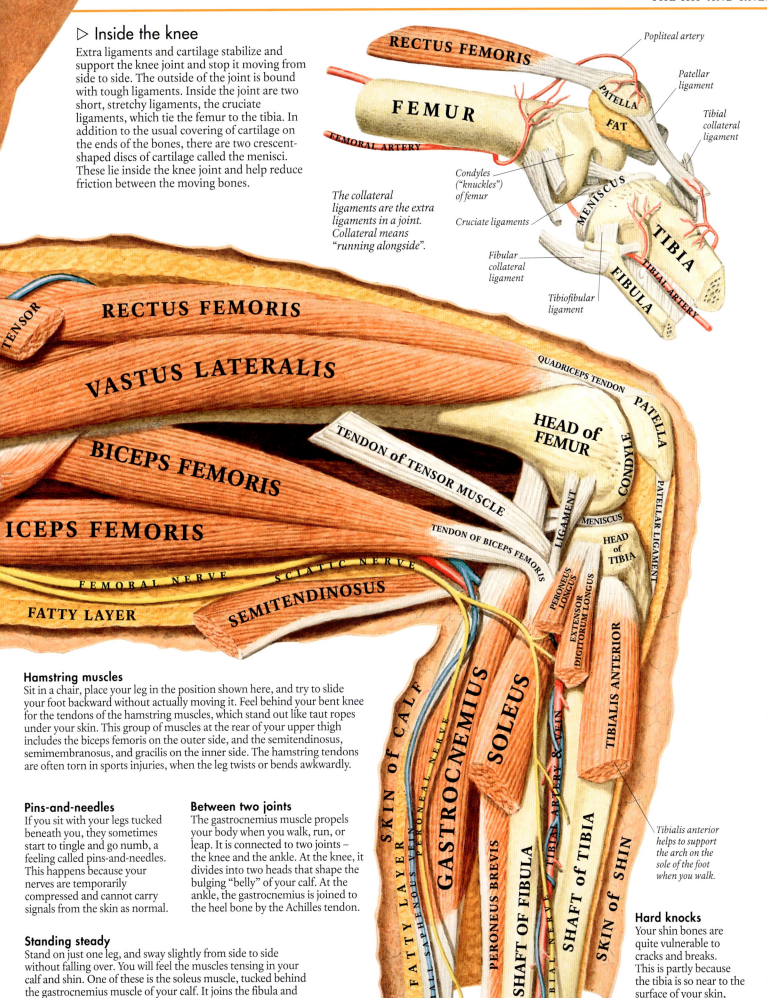

▷ Inside the knee

Extra ligaments and cartilage stabilize and support the knee joint and stop it moving from side to side. The outside of the joint is bound with tough ligaments. Inside the joint are two short, stretchy ligaments, the cruciate ligaments, which tie the femur to the tibia. In addition to the usual covering of cartilage on the ends of the bones, there are two crescent-shaped discs of cartilage called the menisci. These lie inside the knee joint and help reduce friction between the moving bones.

The collateral ligaments are the extra ligaments in a joint. Collateral means "running alongside".

Labels on upper illustration:
RECTUS FEMORIS · Popliteal artery · FEMUR · PATELLA · FAT · Patellar ligament · Tibial collateral ligament · FEMORAL ARTERY · Condyles ("knuckles") of femur · Cruciate ligaments · MENISCUS · TIBIA · Fibular collateral ligament · FIBULA · TIBIAL ARTERY · Tibiofibular ligament

Labels on main illustration:
TENSOR · RECTUS FEMORIS · VASTUS LATERALIS · BICEPS FEMORIS · ICEPS FEMORIS · TENDON of TENSOR MUSCLE · QUADRICEPS TENDON · PATELLA · HEAD of FEMUR · CONDYLE · PATELLAR LIGAMENT · MENISCUS · LIGAMENT · HEAD of TIBIA · TENDON OF BICEPS FEMORIS · FEMORAL NERVE · SCIATIC NERVE · FATTY LAYER · SEMITENDINOSUS · PERONEUS LONGUS · EXTENSOR DIGITORUM LONGUS · TIBIALIS ANTERIOR · SKIN OF CALF · GASTROCNEMIUS · SOLEUS · PERONEUS BREVIS · SHAFT OF FIBULA · TIBIAL ARTERY & VEIN · SHAFT of TIBIA · SKIN of SHIN · FATTY LAYER · SMALL SAPHENOUS VEIN · PERONEAL NERVE · TIBIAL NERVE

Hamstring muscles

Sit in a chair, place your leg in the position shown here, and try to slide your foot backward without actually moving it. Feel behind your bent knee for the tendons of the hamstring muscles, which stand out like taut ropes under your skin. This group of muscles at the rear of your upper thigh includes the biceps femoris on the outer side, and the semitendinosus, semimembranosus, and gracilis on the inner side. The hamstring tendons are often torn in sports injuries, when the leg twists or bends awkwardly.

Pins-and-needles

If you sit with your legs tucked beneath you, they sometimes start to tingle and go numb, a feeling called pins-and-needles. This happens because your nerves are temporarily compressed and cannot carry signals from the skin as normal.

Between two joints

The gastrocnemius muscle propels your body when you walk, run, or leap. It is connected to two joints – the knee and the ankle. At the knee, it divides into two heads that shape the bulging "belly" of your calf. At the ankle, the gastrocnemius is joined to the heel bone by the Achilles tendon.

Tibialis anterior helps to support the arch on the sole of the foot when you walk.

Standing steady

Stand on just one leg, and sway slightly from side to side without falling over. You will feel the muscles tensing in your calf and shin. One of these is the soleus muscle, tucked behind the gastrocnemius muscle of your calf. It joins the fibula and tibia to the heel bone on the foot. This muscle helps to steady the leg when you are standing, by making constant small movements that keep your body balanced over your foot.

Hard knocks

Your shin bones are quite vulnerable to cracks and breaks. This is partly because the tibia is so near to the surface of your skin, with little protective fat and muscle over it.

The Ankle and Foot

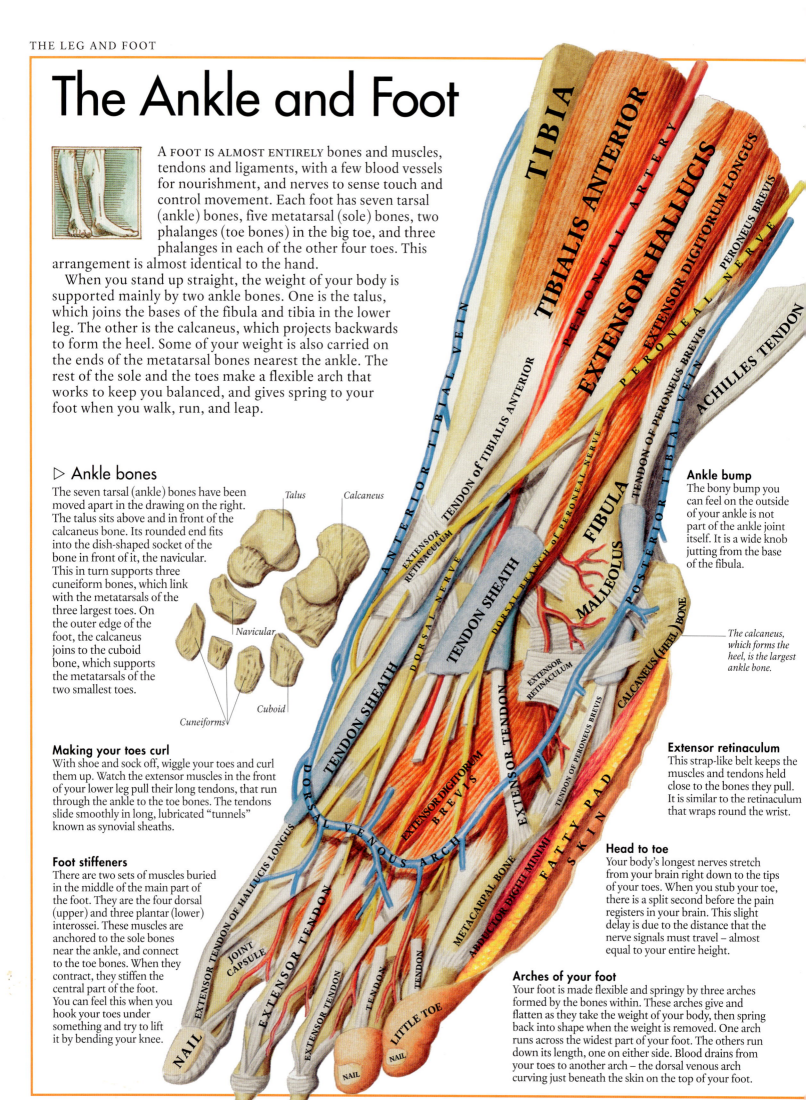

A FOOT IS ALMOST ENTIRELY bones and muscles, tendons and ligaments, with a few blood vessels for nourishment, and nerves to sense touch and control movement. Each foot has seven tarsal (ankle) bones, five metatarsal (sole) bones, two phalanges (toe bones) in the big toe, and three phalanges in each of the other four toes. This arrangement is almost identical to the hand.

When you stand up straight, the weight of your body is supported mainly by two ankle bones. One is the talus, which joins the bases of the fibula and tibia in the lower leg. The other is the calcaneus, which projects backwards to form the heel. Some of your weight is also carried on the ends of the metatarsal bones nearest the ankle. The rest of the sole and the toes make a flexible arch that works to keep you balanced, and gives spring to your foot when you walk, run, and leap.

▷ Ankle bones

The seven tarsal (ankle) bones have been moved apart in the drawing on the right. The talus sits above and in front of the calcaneus bone. Its rounded end fits into the dish-shaped socket of the bone in front of it, the navicular. This in turn supports three cuneiform bones, which link with the metatarsals of the three largest toes. On the outer edge of the foot, the calcaneus joins to the cuboid bone, which supports the metatarsals of the two smallest toes.

Making your toes curl

With shoe and sock off, wiggle your toes and curl them up. Watch the extensor muscles in the front of your lower leg pull their long tendons, that run through the ankle to the toe bones. The tendons slide smoothly in long, lubricated "tunnels" known as synovial sheaths.

Foot stiffeners

There are two sets of muscles buried in the middle of the main part of the foot. They are the four dorsal (upper) and three plantar (lower) interossei. These muscles are anchored to the sole bones near the ankle, and connect to the toe bones. When they contract, they stiffen the central part of the foot. You can feel this when you hook your toes under something and try to lift it by bending your knee.

Ankle bump

The bony bump you can feel on the outside of your ankle is not part of the ankle joint itself. It is a wide knob jutting from the base of the fibula.

The calcaneus, which forms the heel, is the largest ankle bone.

Extensor retinaculum

This strap-like belt keeps the muscles and tendons held close to the bones they pull. It is similar to the retinaculum that wraps round the wrist.

Head to toe

Your body's longest nerves stretch from your brain right down to the tips of your toes. When you stub your toe, there is a split second before the pain registers in your brain. This slight delay is due to the distance that the nerve signals must travel – almost equal to your entire height.

Arches of your foot

Your foot is made flexible and springy by three arches formed by the bones within. These arches give and flatten as they take the weight of your body, then spring back into shape when the weight is removed. One arch runs across the widest part of your foot. The others run down its length, one on either side. Blood drains from your toes to another arch – the dorsal venous arch curving just beneath the skin on the top of your foot.

Labels on the main foot illustration:
LITTLE TOE
FLEXOR TENDON
FOURTH LUMBRICAL
THIRD LUMBRICAL
FLEXOR TENDON
SECOND LUMBRICAL
FLEXOR TENDON
FIRST LUMBRICAL
FLEXOR TENDON of HALLUCIS LONGUS
FLEXOR HALLUCIS BREVIS
FLEXOR TENDON
ABDUCTOR DIGITI MINIMI
FLEXOR DIGITI MINIMI
FLEXOR DIGITORUM BREVIS
PLANTAR APONEUROSIS
FATTY PAD of SOLE
ABDUCTOR HALLUCIS
PLANTAR APONEUROSIS
CALCANEUS (HEEL) BONE

Flexor muscles arch the foot and pull the toes down.

Abductor muscles pull or twist the foot so that the sole faces to the outer side.

At 15 cm (6 in) long, the Achilles tendon is the longest tendon in your body – and the strongest.

ACHILLES TENDON (CALCANEUS TENDON)

Calcaneus tendon
The calcaneus tendon, linking the gastrocnemius and soleus muscles in the calf to the calcaneus bone, is also called the Achilles tendon. Achilles was a hero in Greek mythology. When he was a child, his mother tried to make him immortal by dipping him in the waters of the River Styx. Because she held him by his heel, it never touched the water, making this his one weak spot. Achilles was later killed in battle when an arrow pierced his weak heel.

The fibres of a tendon, shown in this microphotograph, are embedded in the outer surface of the bone.

▽ The base of the body
The sole of your foot bears your entire body weight when you stand still, and withstands a force equivalent to up to five times your body weight when you are on the move. Tough straps and bands of dense, fibrous tissue, such as the plantar aponeurosis, secure the muscles and tendons that move the bones. The skin of the sole can be more than 5 mm (0.2 in) thick, the thickest in the body. Toes – like fingers – have ridged prints.

Sole of your foot
Just under the skin and fat is the plantar aponeurosis, a dense network of criss-crossing collagen fibres that forms a secure base for the foot.

Fibre and fat
The thick pads of fat in the soles have stringy fibres growing through and between them. This design works in the same way that the criss-crossed stitching in a duvet stops the filling from bunching up around the edges. It prevents the shock-cushioning fat from being squeezed out to the sides of the foot.

Sole muscles
The main actions of the sole (plantar) muscles are to arch the whole foot, and to curl the toes downward. The most powerful ones pull on the bones of the big toe, or hallux.

▽ Hands and feet
The bones in your hands and feet share not only the same names, but also the same arrangement, as shown in these skeletal views. The hand bones are thinner and lighter and their joints are more flexible. The foot is designed for weight-bearing, although with practice, people unable to use their hands have learned to write, type, and paint with their toes.

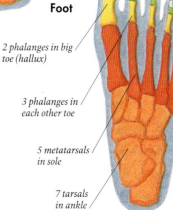

Hand
3 phalanges in each finger
2 phalanges in thumb
5 metacarpals in palm
8 carpals in wrist

Foot
2 phalanges in big toe (hallux)
3 phalanges in each other toe
5 metatarsals in sole
7 tarsals in ankle

GETTING OFF ON THE RIGHT FOOT
No other mammal, and few other animals, can walk with our smooth, two-legged stride. The leg acts as a system of levers. The calcaneus bone of your heel sticks out to the rear of the joint between the shin bones and the ankle. When you take a step, it touches the ground first. Then the calf muscles contract, pulling the heel up. Forward push is provided by the ball of your foot. As your entire body weight shifts to the bones in the front of your foot, its arch shape flattens so that the weight is evenly distributed. One final push is provided by the flexor muscles of your big toe. The front shin muscles lift up the front of the foot, and you are ready to take another step.

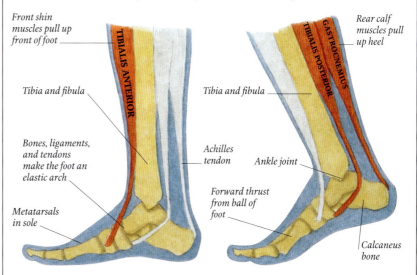

Front shin muscles pull up front of foot
TIBIALIS ANTERIOR
Tibia and fibula
Bones, ligaments, and tendons make the foot an elastic arch
Metatarsals in sole
Rear calf muscles pull up heel
GASTROCNEMIUS
TIBIALIS POSTERIOR
Tibia and fibula
Achilles tendon
Ankle joint
Forward thrust from ball of foot
Calcaneus bone

INDEX

ACKNOWLEDGMENTS

Dorling Kindersley would like to thank the following:
Shelagh Gibson for production; Anna Kunst for translation; Catherine O'Rourke for picture research; Ann Kramer and Miranda Smith for editorial assistance; Lynn Bresler for the index; and Jessica Cawthra for proofreading.

Additional illustrations
Susanna Addario, Jon Rogers, and John Hutchinson

The publisher would like to thank the following for their kind permission to reproduce their photographs:
(Key: a-above, b-below/bottom, c-centre, f-far, l-left, r-right, t-top)

2 Getty Images: Allsport UK / Allsport / Getty Images / Staff (bl). **3 Science Photo Library:** CNRI (ftl, ftr); Simon Fraser (tl); Professors P.M. Motta & J. Van Blerkom (tr). **4 Alamy Stock Photo:** Science Photo Library (bl). **Science Photo Library:** CNRI (bc). **7 Alamy Stock Photo:** Science Photo Library (tr). **10 Science Photo Library:** CNRI (bl, fbl). **14 Alamy Stock Photo:** Science Photo Library (bl). **16 Dreamstime.com:** Heiti Paves (bl). **Science Photo Library:** Medical Imagery Studios / Design Pics (cla). **17 Alamy Stock Photo:** Science Photo Library (cr). **18 Getty Images:** SSPL (cl). **19 Science Photo Library:** Clouds Hill Imaging Ltd (cr). **20 Alamy Stock Photo:** Sebastian Kaulitzki (cra). **22 Science Photo Library:** Simon Fraser. **24 Dreamstime.com:** Iofoto (tr). **Science Photo Library:** Tissuepix (cl). **27 Science Photo Library:** CNRI (tr). **30 Science Photo Library:** CNRI (cl). **31 Science Photo Library:** CNRI (br). **32 Getty Images:** NurPhoto / Contributor (cla). **34 Alamy Stock Photo:** Chronicle (tc). **36 Dreamstime.com:** Flair Images (cl). **38 Alamy Stock Photo:** Science History Images (tc). **Dorling Kindersley:** Department of Cybernetics, University of Reading (c). **39 Dreamstime.com:** Gualtiero Boffi (cr). **41 Science Photo Library:** Kage Mikrofotografie Gbr (tr). **43 Science Photo Library:** Gastrolab (cra); David M. Martin, MD (fcra). **45 Science Photo Library:** CNRI (tr). **46 Science Photo Library:** Steve Gschmeissner. **47 Alamy Stock Photo:** Chronicle (br). **48 Science Photo Library:** Eye Of Science (tr). **51 Science Photo Library:** CNRI (tc). **52 Science Photo Library:** Professors P.M. Motta & J. Van Blerkom (bl). **54 Science Photo Library:** Motta & Familiari / Anatomy Dept. / University "La Sapienza", Rome (cla). **55 123RF.com:** Jovannig (crb). **58 Alamy Stock Photo:** Chronicle (br); Enigma (cl). **60 Dreamstime.com:** Mitchell Gunn (clb). **63 Science Photo Library:** Steve Gschmeissner (cb).

All other images © Dorling Kindersley

For further information see: www.dkimages.com